Pamphlet No. 52

February, 1940

STATE OF IDAHO
C. A. Bottolfsen, Governor

————————————

IDAHO BUREAU OF MINES AND GEOLOGY
A. W. Fahrenwald, Director

————————————

GOLD PLACERS OF THE SECESH BASIN
IDAHO COUNTY, IDAHO

By
Stephen R. Capps

————————————

Prepared in cooperation with
the Geological Survey, U. S. Department of the Interior

————————————

University of Idaho
Moscow, Idaho

CONTENTS

LIST OF ILLUSTRATIONS

GOLD PLACERS OF THE SECESH BASIN
IDAHO COUNTY, IDAHO

By
Stephen R. Capps

INTRODUCTION

This report is the result of one of a series of studies dealing with the gold placer resources of central Idaho, carried out in cooperation between the U. S. Geological Survey and the Idaho Bureau of Mines and Geology. These studies began in 1933 and have continued each year since. They were begun because of the increasing interest in the gold placer deposits of the state, and because it was hoped that an understanding of the conditions under which the gold-bearing gravels had been deposited would be helpful to those engaged in the exploitation of the placers, and indicate where other such deposits might be found.

It has long been recognized that central Idaho has had a complex physiographic history, and that a knowledge of the succession of geologic events throughout this area is necessary if there is to be a proper understanding of the conditions under which the gold-bearing gravels were laid down. The method of approach to this problem that seemed most likely to be fruitful was to study various individual placer camps in detail, and from the conditions found in them to build up a broad picture that would include the major events in the entire province. This procedure has been followed for some years, and has resulted in descriptions published of the placer deposits of the Elk City [1], Florence [2], Warren [3], and Dixie [4] districts, and a general description of the gold-bearing gravels of the Nezperce National Forest. [5] It is apparent from these studies, and from that with which this report is concerned, that the presence of workable gold placers in each of these districts is the result of a series of geological events, some of which are peculiar to the individual district in question, but others of which are common to the region as a whole. The picture is still incomplete, but its larger features are becoming clearer, and further work will fill in the details. There are few references in the literature that relate specifically to the geology or the placer deposits of the Secesh Basin for in such publications as there are it has generally been included with the Warren district. Recently, Lorain and Metzger [6] have published a description of possible placer areas in the Burgdorf district, but have paid little attention to the geologic setting, or to the origin of the placer deposits. Lorain has also recently published an account of lode mining activities in Idaho County. [7]

This report on the gold placers of the Secesh Basin is believed to show the importance of various stages of Pleistocene glaciation in the destruction of pre-

1/Shenon, P. J., and Reed, J. C., Geology and ore deposits of the Orogrande, Buffalo Hump, Elk City, and Tenmile districts, Idaho County, Idaho: U. S. Geol. Survey Circular 9, 1934.

2/Reed, J. C., Geology and ore deposits of the Florence mining district, Idaho County, Idaho: Idaho Bureau of Mines and Geology Pamphlet 46, 1939.

3/Reed, J. C., Geology and ore deposits of the Warren mining district, Idaho County, Idaho: Idaho Bureau of Mines and Geology Pamphlet 45, 1937.

4/Capps, S. R., The Dixie placer district, Idaho, with notes on the lode mines, by Ralph J. Roberts: Idaho Bureau of Mines and Geology Pamphlet 48, 1939.

5/Reed, J. C., Gold-bearing gravels of the Nezperce National Forest, Idaho County Idaho: Idaho Bureau of Mines and Geology Pamphlet 40, 1934.

6/Lorain, S. H., and Metzger, O. H., Reconnaissance of placer mining districts in Idaho County, Idaho: Bur. of Min. I.C. 7023, 1938.

7/Lorain, S. H., Gold mining and milling in Idaho County, Idaho: Bur. of Mines I.C. 7039, 1938.

existing placer deposits, and in the formation of new ones. This area has been an important source of placer gold in past years, although only small scale mining was in progress in 1938. Large areas of ground have been worked, though by somewhat primitive methods. It is believed that considerable ground remains that could be profitably exploited at present, and there is still more ground that may be found workable in the future.

The field season extended from late June to late September, 1938. The principal attention was given to the study of the placer deposits here described, although the larger problem of the physiographic history of central Idaho was kept constantly in mind and reconnaissance was carried into adjacent areas. The writer was ably assisted throughout the entire field season by J. Eric Bucher and Max E. Willard, who carried out the detailed topographic surveys here presented and also contributed to the geologic mapping. Their conscientious efforts are greatly appreciated, and the value of this report is much enhanced by the topographic sketches they made.

The hearty cooperation of the local residents, mining men, and others was a source of great satisfaction. Much of the mining was done in the days before adequate production figures were kept, and the only record of early operations and of the early gold output is that obtained from old residents. Such records are not accurate, according to modern standards, but at least they give a general idea of the dates when the different properties were operated, of the mining methods employed, and of the amount of gold recovered. It is impossible to list all those who furnished information, but especial thanks are due to L. E. Winkler of the Golden Rule mine; to E. F. Stratton, lessor and miner at Ruby Meadows; to James Harris of Burgdorf; and to Theodore Mauersberger, Sam Walker, Ren Wahn, and Mark Evans. During the summer, the writer was favored by short visits from D. F. Hewett and J. T. Pardee of the U. S. Geological Survey, both of whom made helpful suggestions.

Little geologic work had heretofore been done in the Secesh Basin. Ross paid a short visit to the adjacent Warren district in 1930, but his report has not been published. [1] The Warren district was studied in 1935 by J. C. Reed.[2] The gold production of the Secesh Basin has usually been included with that of the Warren district. The Florence district, about 20 miles to the northwest, and the Dixie district, about 26 miles to the northeast, have also been recently studied. These three districts have certain features in common, but each shows individual characteristics that distinguish it from the others. Some scattered references to the Warren district are to be found in such publications as the annual reports of the Idaho State Inspector of Mines, the annual volumes of Mineral Resources of the United States, the Minerals Yearbook, and the technical press, but most of them contain little geologic information and almost nothing about the area here under consideration.

GEOGRAPHY

The location of the Secesh Basin in relation to the State of Idaho as a whole is shown in Figure 1. The placer deposits of the Secesh Basin are scattered along the valley of the Secesh River itself, between the upper end of its canyon and Secesh Summit, and along its tributaries, Grouse, Ruby, and Lake creeks. The northernmost old workings on Lake Creek are 12 miles distant by road from the lowest workings on Secesh River. The upper basin of Secesh River, and the location of the various placer areas covered by detailed maps are shown in Figure 2. No topographic map of this region exists, although the U. S. Geological Survey

[1] Ross, C. P., The Marshall Lake and Warren mining districts, Idaho County, Idaho. Manuscript report.
[2] Reed, J. C., Idaho Bur. Min. and Geol. Pamphlet 45, op. cit., 1937.

2.

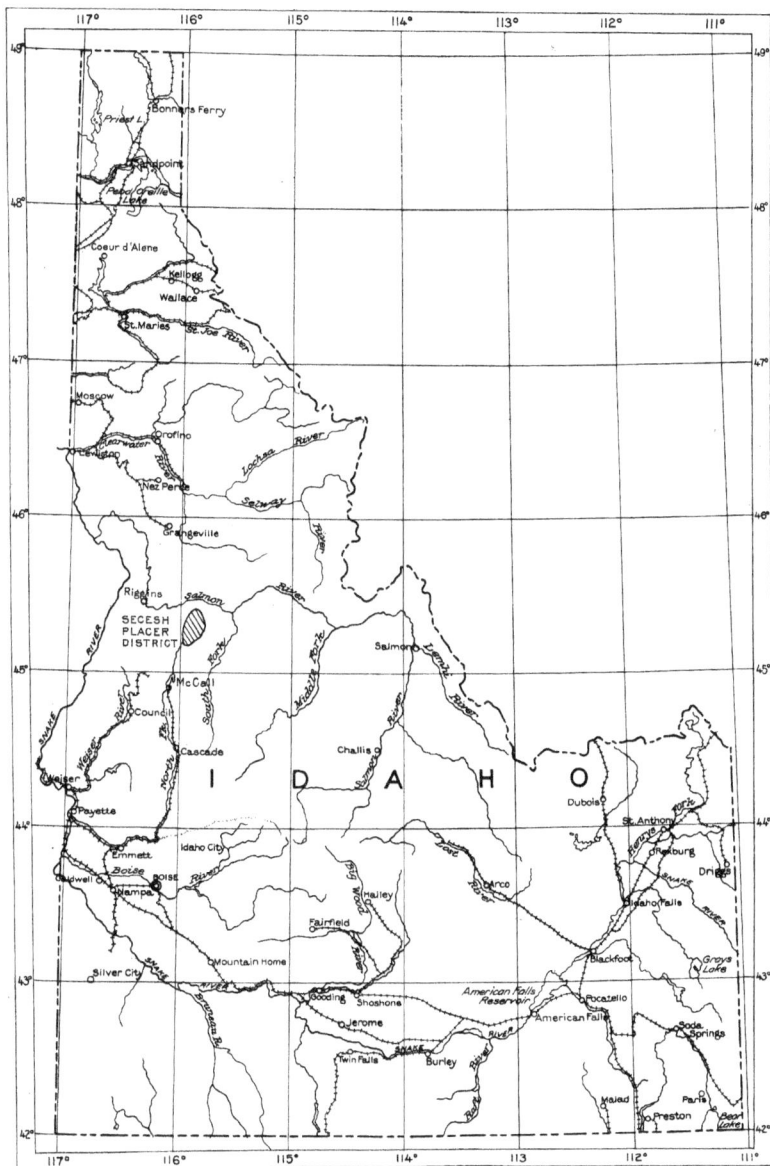

FIG. 1. INDEX MAP OF IDAHO SHOWING LOCATION OF THE
SECESH PLACER DISTRICT.

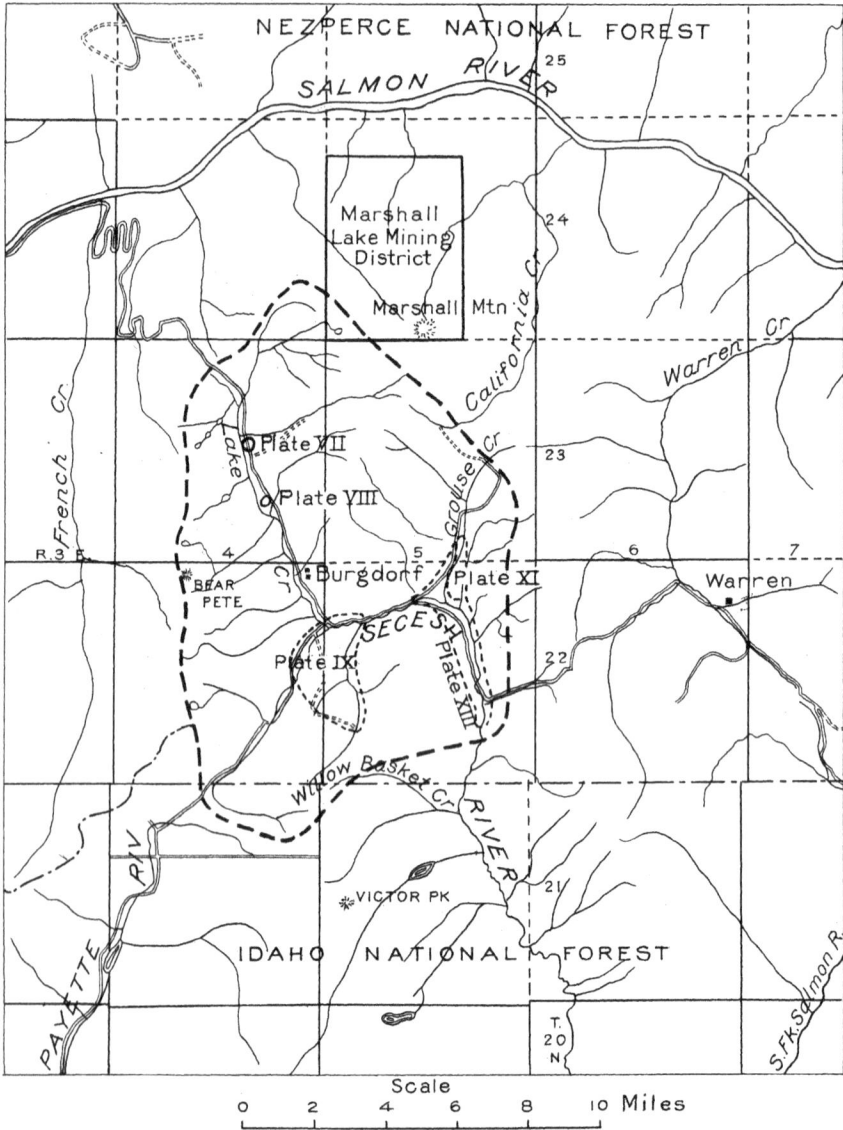

FIG. 2. SKETCH MAP OF UPPER SECESH BASIN.

has under way a project that contemplates mapping in the near future a thirty minute quadrangle to cover this area. Most of the district has been surveyed by the General Land Office, and the section corners are in place. All of the area here considered lies within Townships 22 and 23, N. R. 4 E., and Townships 22 and 23 N. R. 5 E., and lies within the boundaries of the Idaho National Forest in Idaho County, Idaho. In most mining descriptions it has been included in the Warren mining district, although it lies immediately adjacent to, and to the south of, the Marshall Lake district. Inasmuch as time was not available during the 1938 field season for surveying topographically all of this basin, it was thought best to map certain critical areas on as large a scale as possible. Accordingly, a map was prepared of the Secesh Meadows from the head of the Golden Rule placer workings to the head of the canyon, on a scale of 1:24,000 and a contour interval of 25 feet (Plate XIII); a more detailed map that covered most of the Golden Rule ground on a scale of 1:6000 with a contour interval of 10 feet (Plate XI); a map of the placer ground between Ruby Creek and Secesh River on a scale of 1:12,000 and a contour interval of 25 feet (Plate IX); and topographical sketches of two old placer workings on Lake Creek (Plates VII and VIII).

The settlement of Burgdorf, on Lake Creek a mile above its mouth, is centrally located in the area here under discussion, and is the supply point for the district. It has a general store and filling station, a hotel, a swimming pool supplied by hot springs, and a number of cabins for the accommodation of visitors. Burgdorf lies at an altitude of 6,100 feet above sea level, and has a delightful summer climate.

Until comparatively recent years, this district was difficultly accessible. Burgdorf is 31 miles north of McCall, a lumber and resort town on Payette Lake, and the nearest railroad connection, but the Bureau of Public Roads has improved the route from McCall to Warren by a road passing centrally through this portion of the Secesh Basin, and this road has brought Burgdorf to within an hour's travel by car from McCall during the open season. It is possible in midsummer to proceed eastward from Warren by a Forest Service road to Edwardsburg, and thence southward by way of Profile Gap, through Yellow Pine to Landmark and Cascade, although the road from Warren to Edwardsburg is now little used. Within the last five years, road construction has been vigorously carried on by the U. S. Forest Service, and by the Civilian Conservation Corps under the direction of the Forest Service. A new C.C.C. road now leaves the Idaho North-South highway at Riggins, proceeds up the Canyon of Salmon River to French Creek, and then, climbing steeply out of the canyon, extends southward past Burgdorf to join the McCall-Warren road at the mouth of Lake Creek.

Several branch roads from the routes mentioned above are kept open during the summer. One from the Burgdorf-French Creek road runs to the mines of the Marshall Lake district; another leaves the Warren road at the mouth of Grouse Creek and runs to the War Eagle lookout station, with a branch to upper California Creek; and still another leaves the Warren road just above the mouth of Lake Creek and extends to Ruby Creek.

The road from Burgdorf to McCall crosses the Secesh Summit at an altitude of 6,450 feet. The divide between Lake and French creeks is at about the same elevation. All of this high country receives a heavy snow fall, and the roads are commonly closed to automobile travel from the middle of November to the middle of June. During that period, an attempt is made to keep communication open for mail and express between Warren and Burgdorf and McCall by tractor. All of the heavy supplies for the Marshall Lake district, and most of those for Burgdorf and Warren, are brought in by truck during the summer open season.

TOPOGRAPHY AND DRAINAGE

Central Idaho falls within the Northern Rocky Mountains province, and in the latitude of the area here under discussion is an unbroken mass of rugged country across the entire state. South of the Salmon River a part of the range is known as the Salmon River Mountains, and north of that river the term Clearwater Mountains is used, although both of these groups are parts of the same mountain mass, and have had much the same geologic and physiographic history, but happen to be separated by the profound but relatively young canyon of the Salmon River. Although this entire province consists of rugged mountains, it should not be understood that the summit ridges consist of a series of sharp ridges and craggy peaks, for that is not the case. Instead, the summit ridges are for the most part rather rounded with smoothly even crest lines, and are timbered to the top. In the Clearwater Mountains these summit ridges rise to altitudes of about 7,000 feet, but in the Secesh district the loftiest points of the Salmon River Mountains attain heights of a little more than 8,000 feet, and farther to the southeast and east the ridge crests rise still higher. From any point on the ridge tops, one seems to be looking across a great maturely dissected plateau, into which most of the streams have cut their valleys to a depth of 2,000 to 2,500 feet. In a few areas, one or a group of peaks rise some distance above this prevailing summit level, and the Salmon River and many of its tributaries have cut deep, narrow canyons far below the level of the upland stream basins. These canyons are rapidly working back into the upland, but large areas remain that are still unreached by the wave of rejuvenation that is spreading from the deep canyons. Except for these young canyons, therefore, this part of central Idaho consists of an upland province of mature topography, many of the ridge crests falling roughly into an undulating surface at elevations of 7,500 to 8,000 feet or a little more, but with a local relief, from stream valleys to ridge crests of between two and three thousand feet. This upland topography obviously was developed before the erosion of the present canyons of Salmon River and its rejuvenated tributaries, and the high summit level has been considered by a number of writers to represent a Tertiary peneplain that had been elevated a few thousand feet and dissected to that depth before the present canyon of the Salmon River was cut. The physiographic history of the area is discussed more fully in a later portion of this report.

Mountain glaciation has played an important role in sculpturing the higher portions of this region to produce the present land forms. During the last or Wisconsin stage of glaciation, most of the ridges that rose to altitudes of 7,500 feet or more were able to foster local glaciers, and these ice streams moved down the valleys for some distance. The glaciers on the north and east slopes were larger than those that faced to the south or west, and some of the more vigorous ones, especially those that drained into the steep young canyons tributary to the Salmon River, pushed down to altitudes as low as 4,000 feet. In the Secesh district, none reached much below 6,000 feet, and many valleys above that altitude were free of ice. There is evidence that the higher parts of this region were glaciated by at least one, and probably two Pleistocene ice advances earlier than the Wisconsin, and that some of those earlier glaciers were considerably larger and extended down the valleys farther than the Wisconsin glaciers.

This report is concerned with the headward basin of Secesh River, and the basins of its tributaries, Grouse Creek, Ruby Creek, Lake Creek, and their branches. Secesh River, through its headward fork, Summit Creek, rises in a mass of mountains to the east of Secesh Summit, in which the highest peaks are well over 8,000 feet in altitude. Lake Creek, its largest tributary, drains the western slopes of Marshall Mountain, and the eastern slopes of the Bear Pete

Ridge, both of which are more than 8000 feet high. The ridges surrounding upper Grouse Creek also rise above 8000 feet. All of these basins show strong evidence of Wisconsin glaciation, and contain even more extensive morainal deposits of pre-Wisconsin age. Inasmuch as these successive glacial advances have had much to do with the present distribution of the gold placer deposits, they will be more fully discussed in a later section of this report.

The valleys of Upper Secesh River, in the area here treated and of Lake Creek as well, are characterized by alternating stretches of broad, open meadows separated by narrow canyons. This condition is not what one would expect by normal mature erosion of a region underlain by rocks of remarkably uniform composition, but can best be explained as the result of shifts of the stream channels due to stream dislocation during periods of glaciation.

A conspicuous feature of this district is the peculiar arrangement of its drainage pattern. Lake Creek, rising within 5 miles of Salmon River, flows southward directly away from it, and after traversing a circuit nearly 100 miles long reaches Salmon River at a point only a few miles from its source. It is now recognized that this stream's course was determined by structural deformation of the mountain mass, and that Lake Creek lies along the trace of a great fault, as does the Secesh River below Grouse Creek. Grouse Creek itself follows a minor fault. Evidence is accumulating that many streams in the Clearwater and Salmon mountains follow faults or downwarps of the upland that took place after the extrusion of the middle Miocene Columbia River lava flows, and before the earliest recognized stage of Pleistocene glaciation.

CLIMATE

The Secesh district lies in an alpine province and has a climate determined in large part by its altitude. The lowest point in the district, at the lower end of Secesh Meadows, is just below the 5700 foot contour, and the ridges and peaks rise to altitudes of between 7000 and 8400 feet. No continuous weather records have been kept in this district, but records from similarly located stations indicate a mean monthly temperature in midwinter of about 20 degrees, and about 60 degrees in midsummer. The annual precipitation is between 30 and 40 inches, most of which falls as snow during the winter months. Ten feet of snow on the ground at one time is not unusual, and winter temperatures of 30 degrees or more below zero are common.

The summer climate is delightful, with many bright, warm days and cool nights. Frost may be expected during any month of the year. The summers are dry, with only occasional rain coming as thunder storms.

Underground mining can be carried on throughout the year, but hydraulic placer mining is restricted to the summer, and most of such operations are hampered by a shortage of water after mid-July. In the nearby Warren district the largest dredge is operated throughout the year, but the smaller one is closed for three or four months during the severest winter weather.

VEGETATION

By far the most prevalent timber in the district is the lodge pole pine, which grows in thick stands on most slopes up to an altitude of 7000 feet or more (Pl. I). Interspersed with the lodgepole pine there is some Engleman spruce and Douglas fir, though nowhere in sufficient abundance to justify commercial logging. On the very highest ridge there is a scattering of limber pine. The lodgepole pine is, therefore, the only important timber resource of

Plate I. Lodgepole pine timber on road north of Burgdorf.

the district. It grows tall and straight, although few trees exceed 18 inches in diameter. It is much used for mining timbers, for building construction and for firewood, but is too small to be profitably exploited for lumber.

Considerable areas along the valley floors of Lake and Ruby creeks, and Secesh River are occupied by open, grassy meadows which throughout the spring and early summer are marshy. These meadows provide excellent grazing for livestock during the summer season, and some of them are mowed for hay. They are underlain by alluvial gravels that contain some gold, and have been extensively prospected as possible dredging ground, although, so far, without further economic development.

The Upper Secesh Basin is too high to support general farming for there is no assurance that heavy frosts will not occur, even in midsummer. The agricultural possibilities are, therefore, limited to the harvesting of hay and to grazing.

GEOLOGY

The entire area of the Secesh Basin is underlain by the granitic rocks of the Idaho batholith, with minor amounts of ancient metamorphic rocks, among which are quartzite, gneiss, and schist, into which the batholith was intruded. Next younger than the granitic intrusive rocks there are rather small areas of tilted Tertiary sediments, preserved only in places where they have been faulted down into structural depressions in the older rocks, and which now appear at the surface only where still younger deposits have been stripped off by placer mining. Unconformably overlying the Tertiary sediments and the granitic rocks there are extensive deposits of unconsolidated Pleistocene and Recent materials that include moraines of at least two stages of glaciation; terrace gravels of two or three distinct ages that are, in part at least, related to Pleistocene ice advances; and Recent stream gravels, with locally extensive deposits of tailings derived from the placer mines.

QUARTZITE, GNEISS, AND SCHIST

The oldest rocks in the district are the quartzite, gneiss, and schist that formed the host rock into which the granitic materials of the Idaho batholith were intruded. These rocks have already been adequately described by Reed[1] for the adjoining Warren district, and,inasmuch as the rocks there are of essentially the same character as those of the Secesh Basin, Reed's description is here briefly summarized.

In the Secesh Basin,gneiss, schist, and quartzite are present in considerable amounts on War Eagle Mountain and in the neighborhood of Marshall Mountain, and quartzite crops out on the Burgdorf-French Creek road opposite the mouth of Bear Pete Creek. No attempt was made during the present investigation to map the detailed bedrock geology of the district except in those areas shown on the accompanying topographic maps (Plates VII, VIII, IX, XI, and XIII), but it may be stated with assurance that more than 90 per cent of the district is underlain by granitic rocks, and less than 10 per cent by gneiss, schist, and quartzite. These metamorphic rocks are believed to represent great roof pendents or inclusions that were engulfed in the batholith at the time of its intrusion.

Although they occupy only a small area in the Secesh Basin, these metamorphic rocks are widely distributed throughout central Idaho. They are the prevailing bedrock in the adjacent Marshall Lake district, and have considerable extent in

1/Reed, J. C., Idaho Bur. Min. and Geol. Pamphlet 45, op cit., pp. 6-7, 1937.

the Nezperce National Forest to the north, and in the Edwardsburg district to the east. As will be shown later, the quartzites in particular assume more importance in the placer deposits of the Secesh district than their areal distribution would indicate, for in the older placer deposits most of the granitic boulders have disintegrated to arkosic sand, and a large proportion of the boulders that remain for the placer miners to handle are of the resistant quartzite.

The quartzite is a hard, glassy rock that shows the effects of metamorphism in its complete recrystallization, and much of it contains considerable muscovite, as well as minor amounts of zircon, apatite, garnet, and other accessory minerals. Locally, it has been invaded by and impregnated with the granitic rocks of the batholith so that the contacts are gradational, although elsewhere the contacts are sharp.

On War Eagle Mountain, and in the vicinity of Marshall Mountain, some gneiss and schist are associated with the quartzite. These rocks include quartz-sillimanite schist, biotite-sillimanite-garnet gneiss, and diopside-clinozoisite gneiss.

Although these metamorphic rocks are present in relatively small areas in this district, their presence has had an important influence on the accumulation of the gold placer deposits, for there is a marked tendency of the gold lode mines to be clustered around areas in which the metamorphic rocks occur, and it is these lodes that have yielded the gold which is now found in the placers. This relationship of gold placers to areas of metamorphic rocks is by no means exclusive, for productive lodes are known at many places in the batholith where no metamorphic rocks have been found, but prospectors have long realized that the areas of metamorphic rocks are favorable places in which to prospect for lodes of the precious metals.

Age of the metamorphic rocks

It has been definitely shown that the quartzite, gneiss, and schist are older than the granitic rocks of the batholith as they are cut by them and have been affected by contact metamorphism as the result of the intrusion of the batholith. Furthermore, the quartzite, gneiss, and schist had already been deformed and folded before the granitic rocks were intruded, and the granitic materials cut across those earlier folds. Similar rocks in the Nezperce National Forest are probably of the same age and they have there been correlated by Shenon and Reed [1] with the Belt series of rocks of pre-Cambrian age that are widely developed in northern Idaho. No further evidence of their age was found in this investigation.

GRANITIC ROCKS

The Secesh district lies well within the limits of the Idaho batholith, and is entirely underlain by it, although still older metamorphic rocks crop out in small areas and younger Tertiary sedimentary beds overlie it in places. The batholith is composed of quartz monzonite and related rocks, and a typical specimen is of medium grain and light gray in color. In this district, the batholithic rocks are much like those already described in the adjacent Warren district where Reed [2] notes that the rock is porphyritic with moderately numerous feldspar phenocrysts, in a ground mass composed mainly of quartz, biotite, and muscovite. The normal monzonite contains about 40 per cent plagioclase feldspar; 20 per cent

[1] Shenon, P. J., and Reed, J. C., U. S. Geol. Survey Circular 9, op. cit., p.9, 1934.
[2] Reed, J. C., Idaho Bur. Mines and Geol. Pamphlet 45, op. cit., pp.7-8, 1937.

7.

potash feldspar, mainly microcline; 10 per cent of micas, with biotite and muscovite in about equal amounts; and the remainder largely quartz with zircon, apatite,
titanite, magnetite, and epidote as the common accessories. Locally, where the
content of microcline is relatively small, the rock is classed as granodiorite, and
the range in texture and composition include pegmatite, aplite, and various hydro-
thermally altered phases.

Except in those places where glacial or other forms of erosion have removed
the products of disintegration and decomposition, the granitic rocks are deeply
weathered throughout the district. Placer mining excavations and road cuts in
many places show the granitic rocks to be completely disintegrated to depths of at
least 20 or 30 feet (Pl. II). Locally, especially along the crests of ridges
where the removal of loose material has been rapid, there are outcrops where resistant portions of the rock project as boulders of erosion, spires, or pinnacles.

Age of the granitic rocks

There are difficulties in making a precise determination of the age of the
batholithic granitic rocks. They are known to cut rocks of Triassic age, and to
be overlain by Tertiary sediments and by middle Miocene lavas. A sample of uran-
inite from placer gravels in the Warren Meadows, presumably derived from rocks of
the batholith, was used as the basis for the determination of the lead-uranium
(plus thorium) ratio, and the calculation indicated an Upper Cretaceous age. This
confirms the conclusion obtained from other evidence, and adds to confidence in
accepting a Cretaceous age for the batholith.

TERTIARY SEDIMENTARY ROCKS

Distribution and character

After the injection of the granitic rocks of the Idaho batholith into this
region, a very long time elapsed before the next younger rocks, the Tertiary
shales, sands, and lignite, were laid down upon them. During that long interval,
the intruded igneous rock slowly crystallized and cooled at considerable depth beneath the surface, and erosion of the overlying materials continued far enough to
strip off perhaps a thousand feet or more of the batholith cover and to cut deeply
into the batholith itself. This period of erosion was carried so far that a mature
topography was developed over central Idaho with open valleys, rounded ridges,
streams with gentle gradients, and a local relief between valley bottoms and adjacent ridge tops that would be measured in hundreds rather than in thousands of
feet. The detailed configuration of that surface is not known and may never be
worked out, but on it certain basins were the loci of deposition of sediments,
probably in large part derived by erosion of the nearby ridges but, in part, of
volcanic origin. In places, these sediments, mainly clays and sands, accumulated
to a depth of several hundred feet. That the local relief was mild, and that the
stream gradients were gentle, is evinced by the fineness of the sediments themselves for the coarsest of them are coarse sands, clays greatly predominated over
sands, and there were long intervals when vegetation accumulated in marshy low-
lands to form peat, now altered to lignitic coal.

How extensive these sedimentary deposits originally were, we do not know.
They never became more than moderately consolidated, and, upon later uplift and deformation of the region, they were readily attacked by erosion and were removed
much more rapidly than the older resistant granitic and metamorphic rocks. In
modern times, there were no natural exposures of these sedimentary rocks for they
were preserved only in places where they had been faulted down below the stream
levels, or where they had been covered by terrace gravels, or by glacial moraines.

Plate II. Granitic bedrock at Golden Rule placer mine, on Grouse Creek, completely disintegrated to depths of 20 feet or more.

They are now exposed only in a few places where placer mining operations have removed the overburden and uncovered them, although prospect holes have disclosed their presence somewhat more widely.

In the Secesh Basin, Tertiary sedimentary beds have been recognized only in the lower valley of Grouse Creek where they were uncovered by the old placer operations of the Golden Rule placer mine, and in the old Thorp placer workings in Secesh Valley a short distance below the mouth of Grouse Creek. Drill holes and prospect pits indicate that the formation is present beneath the present stream gravels of Secesh River as far downstream as the lower end of the meadows. Wherever their base could be observed, they lie unconformably upon deeply disintegrated granitic rocks, and they dip 20 to 25 degrees to the west or southwest. At the Thorp workings, they are cut off to the southwest by a fault that brings their faulted edge against granitic rocks, and doubtless another fault, or a branch of the same fault, cut off the west edge of the beds on Grouse Creek. A similar relationship was observed at the Warren district, 10 miles east, where beds of the same type, disclosed beneath terrace gravels by placer mining operations, dip steeply westward, and are probably there too faulted down into their present position. In neither the Secesh nor the Warren basin would the presence of Tertiary sedimentary beds have been suspected or known except for the chance that they were overlain by minable gold placer deposits, and this suggests the possibility that they may be present in other basins in central Idaho where they have not yet been recognized. At an exposure in the old cut at the Golden Rule placer on Grouse Creek, the Tertiary sedimentary beds appear at intervals for a maximum distance across the strike of 500 feet, and dip westward from 12 to 28 degrees to disappear beneath an unmined remnant of terrace gravels. Apparently, the exposed stratigraphic section is somewhere between 150 and 200 feet thick, and more may be present. Most of the outcrops are weathered down to clay or sandy clay, but in a small gully the unweathered beds consist of an alternation of blue, gray, and buff clay shale, and soft gray micaceous sandstone, with a number of beds of impure lignite from 2 inches to 2 feet thick, and some thin, rusty, iron-cemented layers a fraction of an inch thick. These sediments lie unconformably beneath the gold-bearing terrace gravels and form the bedrock to which the mining was carried. Their consistent westward dip, without reversal, suggests the presence of a north-south fault along Grouse Creek Valley somewhere west of the area in which mining was done. (Plates III and XI)

At a point 1.1 mile below the bridge at the upper end of the Secesh Meadows, there is an old placer pit known as the Thorp diggings where high terrace gravels were mined. The floor of the pit consists of stratified Tertiary sediments that lie unconformably beneath the terrace gravels, strike N. 40 degrees West parallel to the trend of the valley, and dip 15 degrees Southwest. Neither top nor bottom of the section is there exposed and the beds have disintegrated on the surface from weathering, but apparently a stratigraphic section about 250 feet thick is exposed. These sedimentary beds continue with a uniform monoclinal dip to the northwest as far as the upper edge of the placer pit, above which they disappear beneath high terrace gravels, but a few hundred feet to the northwest the steep valley wall is composed of granitic rocks and the contact is doubtless a fault. A ditch excavated across a part of this section gave a fresh exposure that was measured in detail. This partial section, given below, is believed to be fairly representative of the Tertiary beds at this placer.

Plate III. Tilted Tertiary shale, sandstone, and lignite at
Thorp mine, Secesh River.

	Feet	Inches
Tan clay shale		6
Gray micaceous sand		8
Dark gray, carbonaceous shale and impure lignite		10
Blue clay shale, weathered buff on joints	1	4
Fine gray mica and quartz sand		6
Fine blue, clayey, micaceous sand		9
Blue to buff clay shale		5
Gray, micaceous sand	1	3
Black clay shale and impure lignite	4	4
Buff clay shale		2
Impure lignite		6
Blue clay shale	1	1
Buff clay shale	1	8
Blue-gray shale	1	7
Gray quartz and mica sand		8
Black clay shale and impure lignite	2	6
Gray quartz and mica sand		5
Gray to buff weathering clay shale: Manganese stains on joints	4	4
Sandy micaceous clay shale	2	10
Black carbonaceous clay	2	2
	28	6

The most complete exposure of the Tertiary sediments in this district is also found in the old Thorp placer workings, and about 1000 feet south of the section just described. There, a section shows nearly continuous exposures of Tertiary rocks for a distance of 1600 feet at right angles to the strike, with an average dip of 25 degrees southwest. The basal beds lie on quartz monzonite and at the west edge of the area the sediments are cut off sharply by a fault and dip abruptly against quartz monzonite at the fault plane. The geologic relations are shown in Figure 4. An attempt was made to measure this section so as to convey a rough idea of its lithologic character, although the exposures were not entirely continuous, and weathering had so broken down the beds that it was necessary to dig shallow pits at frequent intervals to determine the character of the unweathered material. Where exposures were good, more detail is given; where poor, the description is generalized. The section is as follows:

	Feet	Inches
Impure lignite	1	6
Arkosic sand	10	
Carbonaceous clay shale		8
Blue arkosic sand	40	
Blue to gray clay shale	3	
Coarse gray arkosic sand	11	
Fine buff to gray sand	9	
Coarse arkosic sand with thin clay shale layers	20	
Lignite	1	2
Coarse sand with clay shale streaks	28	
Carbonaceous clay with thin lignite seams	4	6
Blue clay shale and arkosic sand	15	
Blue to gray clay shale, carbonaceous shale and arkosic sand	37	

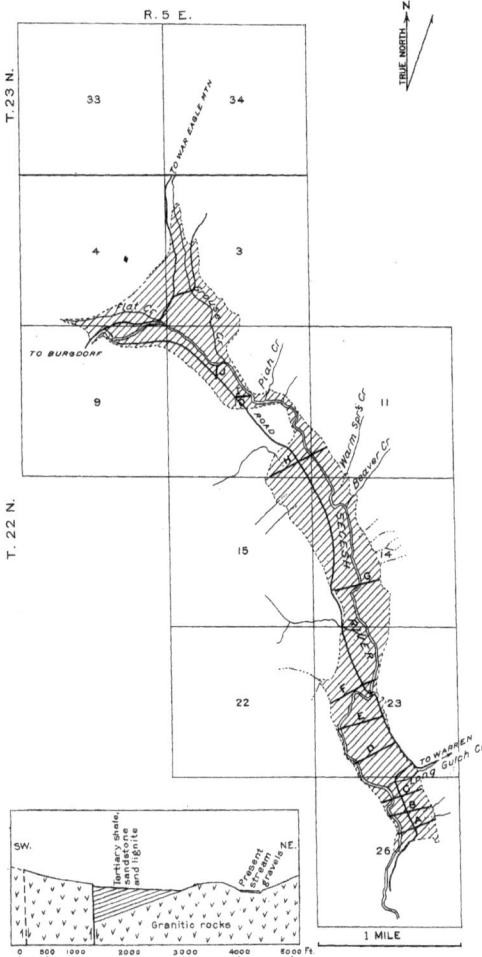

Fig. 4. Structural cross section across Secesh Valley at the Thorp placer, showing tilted Tertiary beds in fault contact with granitic rocks.

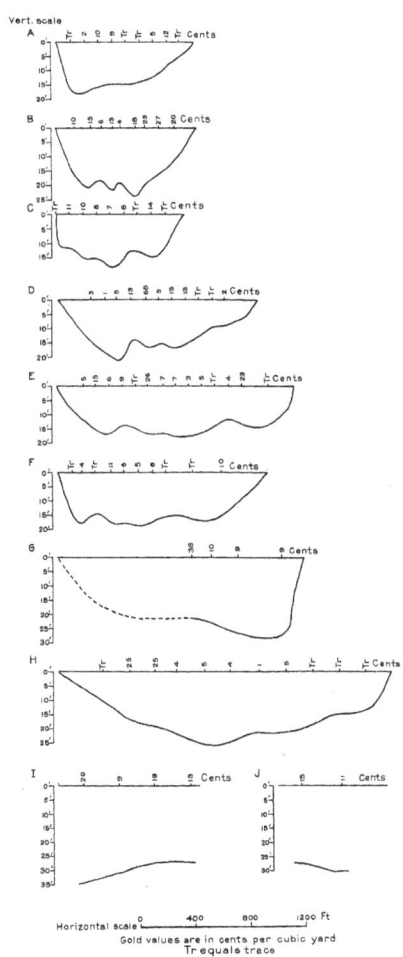

FIGURE 4

FIG. 3. SKETCH MAP OF THE SECESH MEADOWS AND PROFILES SHOWING REPORTED GOLD VALUES.

	Feet	Inches
Thin-bedded, blue and gray clay shale and sand, with several thin lignite seams	60	
Blue clay shale and micacecus sand	24	
Lignite, with shale parting	2	
Arkosic sand	1	
Dirty lignite		5
Blue shale, with 2" lignite seam		6
Lignite		6
Blue shale with lignite seams	1	
Lignite		6
Buff, sandy shale		4
Impure lignite		3
Blue, sandy clay	2	
Buff shale and fine sand	38	
Bony lignite		8
Interbedded blue and gray shale	4	
Cross-bedded arkosic sand	10	
Blue-gray shale with thin lignite seams	9	
Alternating blue and gray shales and sands	52	
Alternating carbonaceous shale and arkosic sands	6	
Hard, mica-bearing, carbonaceous shale	2	
Alternating brown shale and arkosic sand	23	
Sandy shale and micaceous sand	13	
Carbonaceous shale	1	
Blue shale	3	
Micaceous sand	6	
Blue-gray, clay shale and micaceous sand	27	
Fine micaceous sand	23	
Gray and buff, sandy shale and sand	37	
Buff, micaceous, sandy clay	23	
Concealed	100	
Granitic rocks	651	

Structure

As has been shown, sedimentary rocks of Tertiary age are exposed in this district only in the old placer pits of the Golden Rule mine on Grouse Creek and at the old Thorp workings on Secesh River. Their presence is indicated, however, by the logs of borings in the lower Secesh Meadows, and possibly they form the bedrock, now concealed, in the old placer pit 2.2 miles north of Burgdorf. They also have been recognized in one place in the Warren district, although the outcrop exposed by placer mining is now buried beneath dredge tailings. Wherever the structure of these beds could be determined, they have a pronounced dip to the west or northwest, and strike parallel to the valleys in which they lie. In one place in the Thorp pit, they have been cut off on their down-dip margin by a fault, and at other places a similar structure can be inferred on convincing grounds. Doubtless, these rocks were originally deposited over a considerably wider area than that which they now occupy. They are much softer and yield more readily to erosion than the underlying granitic and metamorphic rocks, and they have been preserved only in those places where faulting has dropped them into depressions that are lower than the level at which the streams are now flowing, or or to which the Pleistocene glaciers excavated their valleys. These rocks, therefore, occupy structural valleys formed by faulting. Their presence and attitude is convincing evidence of that faulting, and of the fact that the valleys they occupy are structural, and not merely the product of normal stream and glacial erosion; this proof of the presence of structural valleys in central Idaho greatly strengthens the belief that many other similar valleys in which Tertiary beds have not yet been found are also of structural origin, and has an important bearing on the interpretation of the physiographic history of the central Idaho province.

Practically all of these valleys in west-central Idaho lie along faults that have a north or northwest trend, and in them the block west of the fault has moved upward in relation to the block east of the fault. This relationship is now known to be true in the valleys of Secesh River, Warren Creek, Little Salmon River, Salmon River between Riggins and Whitebird, the Florence Basin and Falls Creek, Meadows Creek, Newsome Creek, and is probably also true of the Elk City and Dixie basins. It may very likely be the reason for the present distribution of many other basins and ridges in central Idaho, the origin of which has not been satisfactorily explained, and for the general physiographic aspect of this region, which is certainly due in an important degree to the prevalence of widely distributed structural valleys.

Age and correlation

The age of the Tertiary sedimentary rocks in this district is not closely known. As will be shown later, they are overlain by deeply decomposed Pleistocene morainal deposits that are much older than the moraines of the Wisconsin stage, and that are probably of early Pleistocene age. Moreover, they occupy structural valleys that had been faulted down to their present position before the early Pleistocene glaciers advanced down them. A few fragmentary fossil plants collected from beds in the old Thorp placer pit have been examined by Roland W. Brown of the Geological Survey, who states that the collection does not contain sufficiently diagnostic material to justify a more specific age determination than "Tertiary". The problem may be approached by another line of reasoning. It is known that most of the block faulting in the western edge of the Clearwater and Salmon mountains took place after the youngest of the Columbia River lava flows had been poured out, for those lavas are displaced by the faulting. The lavas themselves in places have interbedded sediments that carry fossil leaves of the Latah formation, which is of middle or upper Miocene age. The faulting is, therefore, post-Latah and pre-early Pleistocene. It seems fair to assume that the other similar faults in the region are of the same age. Inasmuch as the Tertiary sediments of Secesh Basin can not be definitely correlated with the Latah, though that age for them is here suggested as probable, it can only be stated that these beds are older than the post-Latah faulting, and that they are Tertiary. A closer age assignment is highly desirable as that would possibly also place more accurately the age of the post-Latah faulting, which is of late Miocene or Pliocene age.

PLEISTOCENE AND RECENT DEPOSITS

No deposits have been found in the Secesh Basin area that are intermediate in age between the Tertiary sediments and the oldest glacial deposits of Pleistocene age, yet profound geologic changes took place there in that interval. During middle or upper Miocene time, this mountain province stood at a much lower altitude than at present, or at least was much lower in relation to the lower country to the westward. The mountains then had a mature topography with a relief between adjacent ridges and valleys of perhaps 1000 to 2000 feet, and with streams flowing with moderate gradients through wide, mature valleys. Salmon

12.

River, at that time, probably flowed from east to west across the state along about its present course as far west as the site of ^{R}iggins. Into the lowland west of the mountains there was poured out at that time a tremendous volume of lavas that came as a succession of flows, the aggregate thickness of which locally reached several thousand feet. The basin that received these Miocene lavas was limited on the east in this latitude by the Salmon and Clearwater mountains, then lower than now, with the uppermost flows encroaching somewhat onto the province now occupied by those mountains.

Some time after the last of the Columbia River lava flows had been poured out, the mountain province experienced the period of strong block faulting, previously referred to, presumably in late Miocene or Pliocene times. A general uplift of the whole mountain area of central Idaho accompanied this faulting. As the mountains rose the Salmon River maintained its course across the state by cutting downward through the rising mountain mass, but at the site of Riggins turned abruptly northward to follow along the general course of a great fault. Many of the smaller tributaries of Salmon River also found courses along structural valleys that resulted from faulting. Among these tributaries are Lake Creek, Grouse Creek, and Secesh River below Grouse Creek. The structural basins were modified by stream erosion, which operated upon them during the time between the period of faulting and uplift and the onset of the earliest Pleistocene glaciers. That period was also one of profound erosion by the Salmon River, which cut its gorge across the state to a maximum depth of 6,000 feet. The tributaries, in their efforts to maintain normal gradients to the Salmon, also cut deep canyons, and these canyons pushed back varying distances into the adjacent highlands, their depth, size, and length depending upon the vigor of the streams. In the Secesh River Basin, the newly cut canyon has now reached back just to the foot of Secesh Meadows. Above that point, the valleys have not yet been affected by it.

At least several times during Pleistocene time, a change of climate took place in this region, as well as elsewhere, with increased winter snowfall, and probably also a lowering of the mean annual temperature. Inasmuch as the summit ridges in this region rise to altitudes of more than 8,000 feet, even a moderate cooling of the climate was enough to permit the formations of alpine glaciers in favorably located valleys. These glaciers, at the time of their greatest development, attained considerable size. The one that occupied the upper valley of the Secesh River at one time was 10 miles long. The glacier that moved down Lake Creek from Marshall Mountain had a maximum length of about 12 miles, and an ice stream 8 miles long flowed south from War Eagle Mountain down Grouse Creek and Secesh valleys. The glaciers of the earliest stage that has been recognized here were considerably longer and thicker than those of later stages. As will be shown, glaciers of at least two, and probably three, stages have had a dominant influence on the distribution of placer ground in this district, for practically all of the placer mining that has so far been done here has been on morainal deposits of the glaciers or on terrace gravels that represent outwash from the Pleistocene glaciers. The gravels of the present streams have so far not been worked to any important extent.

Earliest recognized stage of glaciation

A peculiar condition exists in the Secesh Basin in that whereas the granitic rocks of the Idaho batholith comprise perhaps well over 90 per cent of the underlying hard rocks, there are widely distributed deposits of gravel and boulders that are mainly or exclusively of quartzite. Such boulder deposits have been observed along both flanks of Lake Creek Valley to a height of several hundred feet above the valley floor, on the ridge between the Secesh River and Ruby Creek and along the valley sides of Grouse Creek and of Secesh River for more than two

miles below Grouse Creek. These boulder and gravel deposits were of paramount
interest to the placer miners in the early days of the camp for they formed the
gold-bearing placer deposits that were actively exploited. The origin of these
deposits is the source of much speculation and commonly they have been considered
to represent old channel deposits along which gravels were brought in by streams
from some distant source, a not unnatural conclusion in view of the local prevalence of granitic rocks. A close scrutiny of these gravels, however, shows that
everywhere they bear a definite relation to the present valley slopes, and fail
to fall into any pattern that suggests ancient drainage lines not related to the
present drainage. In natural exposures, these gravels appear to have no topographic expression peculiar to themselves, but occur on the surface as scattered
quartzite boulders on normally sloping valley walls. Their surface expression
may be well seen at the settlement of Burgdorf, which is built upon them. Fortunately, the true character of these deposits can now be seen at a number of
places where placer mining operations have opened up good exposures. At present,
the best exposures are found (1) at the old placer workings just above the Burgdorf-French Creek road, at a point 4 miles north of Burgdorf, (2) on the Davis
Mining Company ground between Secesh River and Ruby Creek (Pl. IV), (3) at the
old pit of the Golden Rule placer on Grouse Creek, and (4) at the old Gayhart
Burns pit west of Secesh Meadows (Pl. V). At all of these places it can be seen
that the quartzite boulders are the resistant and undecayed remnants of very old
morainal deposits in which originally the quartzite formed only a small fraction
of the total boulder content, but in which the granitic boulders have now disintegrated to sand. On the surface, the granitic boulders have broken down entirely. At shallow depths, good exposures show the original forms of the granitic boulders, now incoherent aggregates of arkosic materials. At depths as great
as 30 feet, a few of the more resistant granitic boulders are still intact though
much decomposed. In the Warren district, at the placer mine of John Beckner on
Houston Creek, 11 miles northeast of Burgdorf, a similar morainal deposit has
been opened by a cut 80 feet deep in which most of the granitic boulders are disintegrated to that depth, although a few are still hard and intact.

In all of these places the walls of the placer pits, or of more recent
stream-eroded gulleys, cut indiscriminately across boulders and matrix, the
boulders being so rotten that they offer no more resistance to erosion or to the
hydraulic jet than does the sandy interstitial material. The surface configuration of the morainal deposits has been so subdued or altered by erosion that
the glacial origin of the material is no longer evident. The present distribution of this ancient morainal material is, however, such as would be expected if
the glaciers that deposited them had had the same general form and position as
the much more recent glaciers of the Wisconsin stage, the moraines and cirques
of which are now so conspicuous. The earlier glaciers, however, were considerably thicker and reached considerably farther down the valleys than those of the
later Wisconsin stage. Thus, the earlier glacier occupied the valley of Lake
Creek to its mouth, although the Wisconsin glaciers tributary to Lake Creek failed to meet and join in the valley of that stream. The early glacier in upper
Secesh Basin was able to surmount the ridge between that stream and Ruby Creek,
although it probably did not push down Lake Creek much below the mouth of Ruby
Creek. The early glacier that headed in War Eagle Mountain, and the other cirques
at the head of Grouse Creek, pushed down to the mouth of that stream and thence
down Secesh Valley well toward the lower end of the meadows. In all these places
this old morainal material carries some placer gold, and in many localities in
sufficient quantities to have encouraged placer mining in the early days of the
camp. This mining of glacial morainal material for placer gold is unusual, for
most placer deposits result from the concentration of gold by streams from a
large quantity of country rock from which much of the barren material has been
carried on down valley by the streams, leaving the heavy and resistant gold part-

14.

Plate IV. Coarse, thoroughly decomposed moraine of early Pleistocene age, overlain by fresh late Wisconsin moraine (at right). Note tongue of older material that has crept down over the top of the younger moraine. Some six feet of creep zone material have been mined from above both glacial deposits.

Davis Mining Company ground, on Secesh River near mouth of Lake Creek.

Plate V. Early Pleistocene morainal deposit at old Gayhart
Burns mine, west of Secesh Meadows. The granitic boulders are completely disintegrated.

icles behind. Glaciers, by contrast, do not themselves tend to concentrate the gold in their detritus, except where running water has also been effective. The presence of gold in workable quantities in these moraines is probably due to the fact that before the ice advance began there had already been formed concentrations of placer gold in the stream beds, and that this gold and the containing gravels were picked up by the glaciers and incorporated into the moraines.

The positive identification of these deposits as ancient glacial moraines was possible mainly because of the fortunate circumstance that deep, fresh exposures were found at several critical localities. Once identified as ancient moraines, the presence of scattered quartzite boulders in areas underlain by granitic rocks made it feasible to map these old moraines in areas where good exposures were lacking. If, however, there had not been outcrops of quartzite in these glaciated basins, but only granitic rocks of the Idaho batholith, the decomposed remnants of old moraines would in general have been indistinguishable from the arkosic material of weathered granitic rocks in place. The possibility, therefore, presents itself that ancient moraines of similar age may be present in many parts of the mountains of central Idaho, but, as they contain no quartzite or other boulders that are resistant to decomposition, they have so far not been recognized.

Age of early glacial deposits

The pre-Wisconsin moraines just described are obviously older than the moraines left by the last great ice advance of the Wisconsin stage. The granitic boulders of the Wisconsin moraines are hard and firm, the material is generally unoxidized, and the surface forms have been little modified by erosion. Furthermore, at one locality, Wisconsin morainal material was observed to lie unconformably upon an older moraine of much greater age (Pl. IV). In the older moraines, all the granitic boulders have broken down to sand to a depth of 20 feet or more, and in the Warren district a cut shows that most of the granite boulders have disintegrated to a depth of 80 feet, and lie upon deeply decomposed granitic bedrock. The surface features of the older moraines have entirely lost their glacial aspect, and the presence of morainal deposits would scarcely be suspected except for the residual quartzite boulders on their surface. The older moraines extend considerably down valley beyond the outermost stand of the Wisconsin glaciers. Surface creep of the disintegrated upper portion of the old moraines has removed much material, the rotten boulders have been drawn out into thin lenses, and a surficial zone a few feet thick now presents a laminated appearance owing to creep (Pl. VI). Although the glaciers that formed these deposits were in existence long before the last stage of glaciation, they followed valleys that then had essentially the same form and position as the valleys of today.

No accurate correlation of these older moraines with the better known glacial sequence in the Mississippi Valley can be made now, but,after discussion with several geologists long familiar with glacial problems in the western mountains of the United States, all agree that these moraines are at least as old as Illinoian, and they may be as old as early Pleistocene.

Intermediate stage of glaciation

Some evidence found in this district suggests an intermediate stage of glaciation occurred considerably later than the earlier stage, just described, but distinctly earlier than the Wisconsin stage. Along lower Grouse Creek, and along the southwest side of the Secesh Meadows below Grouse Creek, there are high-terrace deposits in which many of the granitic boulders are completely or partly decomposed, although many also are intact and still firm. These terraces stand from 25 to 50 feet above the adjacent stream flats, and represent a valley fill to that elevation in some past period. They contain some placer gold, and have been mined at the Golden Rule placer on Grouse Creek and at the Thorp workings on Secesh River. The level of the terrace surface above the Thorp mine was controlled by the depth of the rock canyon that constricts the Secesh Meadows northeast of the Thorp mine. Below that canyon the terrace surface was graded to the notch at the lower end of the meadows, then not so deep as it is now. The gravels and boulders of these terraces are more advanced in degree of decomposition than are the Wisconsin moraines, or the outwash from them, and their surface stands 30 feet or more above the level of the Wisconsin outwash, yet they are much younger both in degree of decomposition and in their topographic expression than the older moraines nearby. Their presence demands a period of abundant supply of debris to the streams at some time between the period of earliest recognized glaciation and the Wisconsin stage, and suggests the probability of a glacial advance during an intermediate period. However, it must be stated that no morainal deposits of such an intermediate glacial stage have been recognized, although the presence of the high terraces suggests that there may have been one.

<u>Wisconsin glacial stage</u>

The higher mountains of central Idaho everywhere show abundant evidence of a late Pleistocene glacial advance, the Wisconsin stage of glaciation. Practically all of the peaks and ridges that rise to altitudes of 7,000 feet or more had glaciers on their flanks, at least on those north and northeast slopes that were best protected from the direct rays of the sun. From still higher ridges, valley glaciers several miles in length pushed downward into the lower basins. These glaciers have now all melted away, but their former presence is attested by many features characteristic of a glaciated landscape. Steep walled, cirque basins or amphitheaters notch the glaciated peaks and ridges, and in many of them small mountain lakes or tarns occupy basins gouged from the solid bedrock. Below the cirques, one often finds ridge-shaped lateral moraines that slope down stream along the valley walls to merge in the valley below into well-defined terminal morains made up of masses of boulders and rock fragments torn by the ice from the basin above, and dumped in chaotic heaps at the toe of the glacier. In terms of human history, these glaciers of the Wisconsin stage disappeared many thousand years ago, but in terms of earth history their growth and final melting away were comparatively recent events - so recent that post-glacial erosion and rock weathering have done little to alter the appearance of the glaciated basins or of the morainal deposits. In places, the cirque walls and floors still show the polishing and scratching of the bedrock by the over-riding ice; many shallow cirque lakes still show little post-glacial filling by stream-brought debris, and the moraines still have much the same irregular forms that they had when left by the glaciers and the materials of which they are composed are little weathered. In all of these respects the characteristics of the last, or Wisconsin, glacial valleys and moraines are in sharp contrast to those of the earlier, more extensive period of glaciation, already described. As was to be expected, the Wisconsin glaciers overrode and destroyed the deposits of the earlier glaciers in the areas that they covered, but, being smaller, they left their terminal moraines well inside those of their predecessors.

The Secesh Basin, here under consideration, was occupied in Wisconsin times by a number of glaciers, some of which reached a length of several miles. One such glacier had its source near Diamond Rock, at the head of Summit Creek, moved down that stream into the head of Secesh River, and after being joined by a number of tributary glaciers, notably one from Josephine Creek Basin, pushed northward to terminate at about the mouth of Lake Creek. Another glacier, rising on War Eagle Mountain and against the high ridge between the basins of Grouse Creek and Lake Creek, pushed southward about to the upper edge of the Golden Rule placer ground. The main valley of Lake Creek was never ice-filled during Wisconsin time.

16.

although it had been by the earlier **glacier.** A small glacier flowed westward from Marshall Mountain to and below **Marshall Lake,** and,from several basins on the east slope of the ridge of which Bear Pete **Mountain** is the highest point, glaciers moved down to the valley of Lake Creek, but they terminated at the valley edge without flowing southward in to the main valley. The glacier in Pete Creek, rising in a steep-walled cirque against Bear Pete Mountain, pushed down to the eastward far enough to force Lake Creek against its east valley wall, and left a great terminal moraine in Lake Creek Valley that separates the meadows of Lake Creek below Pete Creek from those above. Burgdorf and Nethker Creek basins were also occupied by glaciers that terminated at the west edge of the Lake Creek meadows.

It is here worthy of note that even on a ridge as high as that of which Bear Pete Mountain is the summit, at an altitude of about 8,000 feet, the glaciers on the east slope were vigorous and excavated deep cirques that cut back in the ridge and pushed the divide westward whereas the westward-facing glaciers were feeble and failed to develop well-marked cirques. The same difference in the severity of glaciation on the east slope as compared with the west slope is apparent generally in this district. The ridge between Lake Creek and Upper Grouse Creek is another conspicuous example. It probably resulted not only from the better protection of the east-facing valleys from the direct rays of the afternoon sun, but was due also to prevailing westerly winds that blew the snow across the ridges and allowed it to accumulate in the valleys on the lee slope.

Associated with the Wisconsin moraines, and extending down valley below them are outwash trains of gravel, supplied by the glacier-fed streams and distributed by them in the valleys beyond the limits to which the Wisconsin ice reached. These latest glacial outwash gravels have been little dissected by the streams that deposited them, and in places merge imperceptibly with the present flood-plain deposits. Secesh River, from the mouth of Lake Creek to the head of the Secesh Meadows, has intrenched itself to a depth of 10 or 15 feet into the outwash gravels, but in that stretch flows in a narrow trench, little wider than the stream itself. In places in the Secesh Meadows there are low terraces only a few feet high, but at other places no distinction can be recognized between the Wisconsin outwash gravels and the present flood plain of the stream. The same is true of Lake Creek, and the amount of post-glacial cutting into the valley fill of these streams is, in general, only a few feet.

Present stream deposits

As has been stated, the present stream deposits merge into, and in places are not separable from, the gravels that were laid down as outwash from the last glaciers. Since the final disappearance of the glaciers, erosion by streams has continued, resulting in removal of moderate amounts of material from some of the glacial terraces and in deposition elsewhere. Within the areas occupied by Wisconsin glaciers there has been a tendency to smooth out inequalities in the stream profiles, and to fill in depressions, but that this process is still incomplete is witnessed by the continued existence of many small lakes or tarns. In short, the work accomplished by the streams in post-glacial times is meager. The base level of erosion in this district is controlled by the point, at the lower end of Secesh Meadows, to which the canyon-cutting of the rejuvenated Secesh River has reached in response to the lowering of the Salmon River Canyon, the master stream. It so happens that this canyon-cutting has barely reached the foot of the meadows. In terms of geologic time, however, the meadows will be short-lived, for,once the canyon works back into them,erosion of the valley fill will be rapid, the river will intrench itself in a sharp gorge through the meadows and the wave of rejuvenation will work up the main stream and then up its tributaries.

During the days of vigorous placer mining in this district, the removal of material was greatly accelerated above the normal rate by the hydraulic operations at the placer mines. In Lake Creek Basin, the fine materials were carried down through the meadows and in part deposited there. The operations on the west slope of the hill in the Ruby Creek area resulted in the building of an extensive gravel and sand flat just below the workings, although great quantities of fines were carried down Secesh River. Mining at the head of Ruby Meadows aggraded the flat to a depth of many feet, although the operations farther down Ruby Creek discharged their tailings into a narrow canyon through which most of the material was carried down to the Secesh River. The extensive old operations of the Golden Rule mine on lower Grouse Creek removed about a million and a quarter yards of material, and the fine tailings from the mine have deeply buried the flats of lower Grouse Creek. In the lower Secesh Meadows also the sands from the old placer tailings can be recognized in cut banks along the stream, and a considerable alluvial fan has been built out upon the meadows by the debris sluiced from the old Gayhart Burns mine. All of these accumulations of tailings in the stream flats have buried ground that contains some placer gold, and if these flats are ever mined the cost will be increased because of the necessity for rehandling the barren tailings.

Peat and volcanic ash

Post-glacial deposits in this district include, in addition to the present stream gravels, considerable amounts of peat that represent accumulations of plant remains in marshy or poorly drained areas. These peaty deposits are even now in process of formation over considerable stretches of meadow land on Secesh River, and in the meadows of Lake and Ruby creeks, and their removal is a source of added expense to the placer miner as the peat is tough and felty and is cut with difficulty by the hydraulic stream. This is particularly true where logs are embedded in the peat, as was the case in the ground mined by dragline on Ruby Creek. In exposures made by placer mining on Ruby Creek and its tributary Roland Gulch, at the upper workings on the Golden Rule placer on Grouse Creek, at the Thorp diggings in Secesh Meadows, and at numerous cuts along the streams that flow through the different meadows in the district, the peat is found to be from one to six feet thick. Here and there within it there are thin beds of sand, and some exposures show as many as three interbedded layers of volcanic ash. An exposure at the Golden Rule placer mine gave the following section:

	Feet	Inches
Coarse pebbly sand	3	
Clayey peat	1	3
Fine, white, volcanic ash		1
Peaty clay		2
Gray, volcanic ash		3
Arkosic sand		6
Fine, white, volcanic ash		0.5
Peaty clay		3
Fine sand		1
Bouldery gravel, Wisconsin outwash	10	
Base of exposure	15	7.5

Erosion by creep

The importance of soil creep, or the down slope movement of loose, surface material under the influence of gravity, aided by frost heave, burrowing animals, insects, and worms, the swaying of trees, or any other agencies that cause soil particles to move slightly, has long been recognized, but the rate of such movement is in most places difficult to determine, as is the depth to which creep is effective. In the Secesh Basin, a fortunate combination of excellent exposures

18.

and soil texture gives an unusual opportunity to evaluate the importance of creep as an agency of erosion in this region.

In the placer workings on the Davis Mining Company grounds, 4,000 feet south-southeast of the mouth of Lake Creek, good, cleanly-exposed banks can be examined that cut through the zone of surface creep into the underlying moraines of both the Wisconsin stage and the much earlier stage of Pleistocene glaciation, and in places into the underlying granitic bedrock. There the fresh Wisconsin morainal material is little oxidized and includes great quantities of fresh, hard, granitic boulders, many of which are several feet in diameter, and also a small percentage of quartzite boulders. Above the upper margin of the Wisconsin moraine, and extending to and over the top of the Secesh River-Ruby Creek divide, is an extensive deposit of pre-Wisconsin morainal material (Pl. IX) in which the boulders originally were dominantly granitic, although it also contained some quartzite boulders. In fresh banks there the granitic boulders of this older moraine are now completely disintegrated to arkosic sand, although still retaining their original form and texture. This material is also somewhat oxidized to buff tints, in contrast to the blue-gray color of the Wisconsin moraine. This older morainal material, except for the few remaining quartzite boulders, now responds to creep as though it were a mass of coarse sand, and the distortion of the sands of individual boulders gives evidence of the direction, character, and depth of the creep movements. Plate VI illustrates one of these cuts. In the center of the picture the bank is 15 feet high. For the lower 9 feet, one sees what is obviously an old boulder bed, although the granitic boulders are all so rotten that a pick can easily be driven deeply into them. At a depth of about 6 feet below the surface, distortion of the boulders can be noticed, each being drawn out into a lenticular shape down slope. The surface slope here is about 12 degrees. Nearer the surface the deposit is laminated, but close inspection enables one to identify thin lenses, now several feet long and about an inch thick, that originally were individual boulders. The surficial layer, 1 to 1-1/2 feet thick, has been so disturbed by roots and burrowing animals that it has lost its laminated structure.

The thinness of the laminae near the surface of the creep zone is evidence that the down-slope movement is most rapid near the surface, and becomes progressively slower with depth to a distance of about 6 feet from the surface below which creep is not noticeable. Many exposures were observed on slopes that range in declivity from 5 degrees to 20 degrees, yet irrespective of slope the depth of the creep zone was in all places found to be about 6 feet. An attempt was made to determine whether or not there was a concentration of the resistant quartzite boulders in the creep zone, but, inasmuch as the quartzites were irregularly distributed in the undisturbed moraine and it was difficult to determine an average ratio of quartzites to granites, no definite conclusion could be reached except that there was no pronounced concentration of quartzites in the creep zone.

In the early years of mining in this camp, the miners sluiced large quantities of the creep zone materials, but failed to mine deeply into the underlying and undisturbed old moraines. This preferential mining of the creep zone indicates that there was some concentration of gold in it, owing perhaps to the removal of fine material by rill and sheet wash, and by wind.

At one locality where, unfortunately, most of the creep zone had been removed by mining, an excellent exposure shows Wisconsin morainal material that lies directly upon a deeply decayed older moraine (Plate IV) which extends up slopes above it. The large boulders of the old moraine are oxidized and completely disintegrated. The Wisconsin moraine is unoxidized and the granitic boulders are hard and fresh in contrast to those of the old moraine. In Plate IV, a remnant of the creep zone is shown lapping down over the Wisconsin moraine, but most of the creep zone has been removed by mining. There is evidence from old tailing piles that before any

19.

Plate VI. Creep zone above disintegrated early Pleistocene moraine. Note that individual boulders, now incoherent sand, are drawn out by creep into thin lenses.

mining had been conducted on this ground the creep zone of material from the older moraine had moved down slope about 400 feet over the surface of the Wisconsin moraine. The early miners were interested in sluicing only this crept material and their excavations were carried down only to the surface of the underlying Wisconsin moraine, which was too full of granitic boulders to be profitably handled by the methods then in use.

It seems to be generally agreed that the Wisconsin glaciers persisted in these mountains to a date much later than that usually accepted as the end of Pleistocene time in the central Mississippi Valley region, perhaps about 30,000 years ago. This figure is admittedly only a rough approximation, but if it is of the proper order of magnitude it suggests that creep has accomplished the movement of a sheet of debris some 6 feet thick to a distance of at least 600 feet in three hundred centuries. How much creep material was removed by stream erosion from the base of the slope is not known. As most of this part of central Idaho is underlain by granitic rocks that are generally disintegrated to a depth of many feet, and as it is a region in which most of the surface has slopes of 10 degrees or more, there is reason to think that creep operates over much of the area, and that it is the dominant agency in moving the products of rock weathering down slope into a position where they can be attacked by the streams.

GLACIAL AND POST-GLACIAL DRAINAGE CHANGES

It has not yet been possible to outline the drainage pattern in this region in early Tertiary time when the mountains, then considerably lower than they are now, were reduced to mature slopes and before the extensive lava flows of the Columbia River Basin had spread out. We do know that the extrusion of the Columbia River basalts must have caused great drainage changes in the region directly affected, and must thus have affected also the gradients of the streams that flowed from the mountains. More far-reaching changes in the drainage pattern within the mountains were, however, brought about by the post-lava uplift of the mountain area, and by extensive block faulting in it. The trend of these faults was in general north or northwest, and the subsidiary streams were forced to follow valleys formed by the block faults, although it is believed that the master stream, the Salmon River, was able to maintain its earlier course in a westerly direction across the mountains by cutting across the faulted blocks as the faulting took place. As stated earlier, this period of mountain uplift and faulting occurred some time between the termination of the period of lava extrusion in this latitude and the earliest Pleistocene glaciation of which we have record, and probably was of Pliocene age. Certainly, by early or middle Pleistocene time the stream pattern was much as it is now for the earliest recognized glaciers followed valleys that were much like those now existing, and by that time the canyon of Salmon River had already been deeply incised, although not quite to its present depth.

In the Secesh district, by early Pleistocene time one prominent structural depression extended southeastward from the head of Lake Creek to the canyon of Secesh River. Its position is marked by the course of Lake Creek, the basin of Ruby Creek, the lower valley of Willow Basket Creek, and probably also by the lower canyon of Secesh River. Another structural basin is marked by the course of the Secesh River from Flat Creek at least as far down stream as the foot of the meadows. These structural valleys were then drained by streams that had much the same pattern as those of today.

During the earliest recognized period of Pleistocene glaciation, an ice tongue occupied the valley of Lake Creek as far southward as its mouth. There the ice itself, or the moraines it deposited, blocked the earlier course of that stream and forced the draining waters to find an outlet across a granitic ridge east of the axis of the valley. A canyon was cut through this rock spur, and with the

20.

retreat of the glacier Lake Creek continued to occupy that canyon as it does now.
As one leaves the main Warren road to turn north to Burgdorf the road follows this
narrow canyon for a short distance to emerge above into the wide, flat meadows of
lower Lake Creek. The course of Lake Creek through this canyon was established
during an early stage of Pleistocene glaciation, and has been maintained from that
time through the Wisconsin stage and to the present.

Another Pleistocene drainage change occurred in the Secesh Valley a half mile
below the mouth of Grouse Creek where Secesh River now flows for about 2,000 feet
through a relatively narrow rock canyon that separates the lower Secesh Meadows
from the Secesh-Grouse Creek flats above. At some time between the earliest recognized glacial stage and the Wisconsin stage there was a great outpouring of
gravels into Secesh Valley, probably as a result of an intermediate stage of glaciation, the moraines of which have not been found. These gravels filled the upper
basin to a depth of 30 feet or more, and permitted Secesh River to find a channel
across a gap in a granite spur that projected into the valley from the northeast.
Secesh River has persisted in that channel ever since, and the remnants of the
old gravel fill above the canyon now remain as conspicuous terraces. The outwash
from the Wisconsin glaciers was carried down through this canyon, and in the flat
above the canyon the Wisconsin outwash has been trenched to a depth of only a few
feet.

REGIONAL PHYSIOGRAPHIC HISTORY

Some mention has been made in the preceding pages of the physiographic processes that have resulted in the present distribution of the geologic units described, particularly the glacial and post-glacial deposits. At the risk of some
repetition, it is thought worth while to discuss here the sequence of geologic
events in central Idaho and their influence upon the present surface forms, for
these events have had a direct bearing upon the formation of the gold placers of
the region. Little is known about the appearance of the landscape in early Mes-
ozoic time, but the development of the land forms as we see them today began with
the intrusion of the Idaho batholith, probably in Upper Cretaceous time. The injection of this great mass of material into the upper crust presumably elevated
the surface into a broad dome, and this upland underwent a long period of erosion,
during which several thousand feet of cover were worn away, and the granitic rocks
exposed over wide areas. This old land surface was reduced by erosion to a mature
rolling topography with a relief between adjacent valleys and ridges of perhaps
1,000 feet, and with certain unreduced areas that stood considerably above the
average altitude of the intervening ridges. This old surface has been frequently
referred to as the Idaho peneplain, and, indeed, as viewed from many points on its
surface, there is a striking accordance in elevation of the higher ridges, although between these ridges there are considerable areas in which no summit reaches the skyline. In the Salmon River Mountains, south of Salmon River, the assem-
blage of ridges that together give the impression of an ancient, rather level surface of erosion, is found upon closer examination to consist of ridges whose tops
are rounded, with few flat or plateau-like remnants to indicate that that surface
ever approached a plain. The only such area noted is a flat an acre or so in area
on top of War Eagle Mountain. In the Clearwater Mountains, north of Salmon River,
however, there are considerable areas, such as Columbia Ridge and the high surface
between Buffalo Hump and Gospel Peak, that conform to an old erosion surface.
They are by no means level, but indicate that the ancient surface was mature.
Buffalo Hump and Gospel Peak rise conspicuously above that surface. When it was
being formed, the base level of erosion was much lower than it is now, and the
maximum relief in the region probably was about 2,000 feet. This old high surface
has since been elevated. Its present altitude is between 7,000 and 8,300 feet in
the western part of the state and appears to rise to the east and southeast.

No general agreement has yet been reached as to the period during which this old erosion surface was developed. Evidence of ancient peneplanation, or of the development of mature erosion surfaces, has been obtained at many places in the western mountains, but it is hazardous to infer that separate mountain masses, even though not far apart, have had the same physiographic history, and the correlation of erosion surfaces from one region to another presents many difficulties. This problem has been discussed rather fully by Mansfield [1], who cites the literature that deals with this general region. In his own extensive work in southeastern Idaho he recognizes a succession of erosion surfaces that date from Eocene time to the present. Nothing like so complete or so complex a history can yet be outlined for the province here discussed. In the Buffalo Hump district in the Clearwater Mountains to the north, there are two well developed erosion surfaces below the summit-level surface and above the present canyons of the Salmon and the South Fork of the Clearwater River. Apparently, the uppermost flows of the Columbia River basalt were extruded upon a surface cut 1,000 feet below the summit level. If this statement is true, then the higher surface was developed and later elevated and dissected, giving way to the lower surface by middle Miocene times. Neither of these surfaces below summit-level has been recognized in the Secesh district. In discussing this problem, Ross and Anderson [2] agree that the summit erosion surface in central Idaho cuts volcanics of Miocene age, but is older than the Columbia River basalt flows.

The earliest well established datum in this region is the period of extrusion of the Columbia River basalts that are of middle or upper Miocene age. The next younger geologic events that can be dated with reasonable accuracy are the two and possibly three stages of Pleistocene glaciation. In the interval between the upper Miocene and early or middle Pleistocene, the region was subjected to mountain building movements that included extensive block faulting and warping as well as regional uplift of the Salmon and Clearwater mountains. The block faulting and warping brought into existence many structural valleys that trended in a north or a northwest direction, and the minor drainage was directed along these structural troughs. It is believed that before this the Salmon River had followed about its present westerly course across the state but that this vigorous stream was able to intrench its valley as rapidly as the mountains arose, and so was enabled to maintain its course as far westward as Riggins where it too was diverted northward along a fault valley. This conclusion is somewhat at variance with the belief of some earlier writers who, not having recognized the general prevalence of structural valleys in the region, had attributed the general northerly or northwesterly trend of the minor drainage to the schistosity of the bedrock, or to the alignment of the rock components in those directions. Schistosity and the trend of the geologic units may have had some influence on the drainage pattern, but they were subordinate to faulting and warping. In the district here under discussion, one sees the curious anomaly of Lake Creek, rising in a divide only 5 miles south of the Salmon yet flowing along structural valleys south and southeast away from that river by way of Secesh River to join the South Fork of the Salmon River, and thence in a great loop doubling back on itself so that the water travels nearly 100 miles to cover an air-line distance of only 5 miles.

Such drainage anomalies as this have sometimes been interpreted as indicating that during the mature erosion of the upland surface the master streams of the region also had north or south courses, but no positions of those courses have been recognized. It has been suggested also that the present westerly course of

1/Mansfield, G. R., Geography, geology, and mineral resources of part of southeastern Idaho: U. S. Geol. Survey Prof. Paper 152, p. 11, 1927.
2/Ross, C. P., Mansfield, G. R., and Anderson, A. L., Erosion surfaces in Idaho, Discussion: Jour. Geol., vol. 38, No. 7, pp. 643-657, 1930.

the Salmon River across the state was established rather recently and took place by superposition just before the beginning of the erosion of the present canyon. The present report is not the place to discuss these interpretations fully, nor is our present knowledge of the region adequate to solve the problem of drainage changes in its entirety, yet it is pertinent to say here that accumulating evidence tends to show that much of the drainage pattern of central Idaho is the result of block faulting and warping in post-middle Miocene times. Many of the streams in the high country roughly follow late Tertiary faults or warped troughs, and their courses have been determined by such structural dislocations and not as the result of mature erosion on a surface of diverse rocks.

After faulting, warping, and uplift had ceased, there followed a period during which the streams were engaged in adjusting their courses and gradients. Lake and stream deposits were laid down in the basins formed by faulting and warping, and in places a third erosion surface was developed below the older two that are partly preserved in the Buffalo Hump district. In the mean time, the major streams of the region, notably the Salmon and Clearwater rivers, had been rejuvenated, in part as a result of the increased elevation of the mountains and in part by the development of deep canyons along the master streams, the Snake and Columbia rivers, and had incised themselves deeply into their present canyons. This canyon-cutting brought about a wave of rejuvenation that affected all the tributary streams likewise, and they pushed their canyons constantly farther headward.

Just when in late Miocene or early Pliocene time the period of mountain uplift, faulting, and warping occurred is not known, but certainly it occurred so early that by the end of the Pliocene the canyon of the Salmon River had been incised well toward its present depth, and likewise the tributary streams had pushed their canyons back into the uplands to somewhere near their present length.

The end of Pliocene and the beginning of Pleistocene time was marked by a climatic change that brought about vigorous glaciation in the northern part of the northern hemisphere so that vast areas that earlier had supported a vigorous growth of vegetation were invaded by glacial ice. The time of the first period of glaciation in these mountains has not been determined from evidence obtained there, but it seems reasonable to presume that it occurred at the same time as the first glacial stage elsewhere on the continent. The record is incomplete as to the number and the correlation of the glacial stages in Idaho, but two distinct stages have been recognized, one that certainly occurred in early or middle Pleistocene time, the other in the last or Wisconsin stage. There is also some indirect evidence of one more stage which intervened between these two.

The end of the Wisconsin stage of glaciation in this district is so recent an event geologically that the aspect of the country has changed little since. The canyons have been somewhat further deepened, but not conspicuously. The Wisconsin glacial moraines are still fresh, little oxidized, and not noticeably altered by erosion. The cirque basins still have steep rock walls and many glacial lakes in them show little filling. Perhaps the greater part of post-glacial erosion has been accomplished by soil creep which is constantly moving great quantities of rock waste down the slopes to deliver it to the streams.

The important role of soil creep in the removal of rock debris in these mountains suggests that caution should be exercised in interpreting remnants of erosion surfaces as representing actual land surfaces of great age. It has been shown (pp. 18-20) that even since Wisconsin time notable amounts of comminuted rock materials have been moved by creep, with a consequent reduction of the surfaces from which the debris came. Yet the high-level erosion surface of central Idaho has been interpreted by some as being the actual survivor of an early

Tertiary landscape. Even casual observation shows that most of the granitic ridges of the Idaho batholith, where not scoured by the Wisconsin glaciers, are deeply disintegrated to depths of as much as 20 or 30 feet, and these ridges are generally narrow-crested with slopes of 10 to 20 degrees or more toward the bordering valleys. On such slopes that have thick surface layers of arkosic material, creep is very active. Under these conditions, no ridge top can long maintain its altitude. Lowering is constant and relatively rapid. Even since mid or early Pleistocene time the old morainal deposits have entirely lost their morainal topographic form and have been greatly reduced in thickness. The amount of lowering of such a ridge top by creep since early Tertiary time would certainly be measured in tens, and possibly in hundreds of feet. From these facts, one must conclude that the old erosion surfaces that we now recognize are not remnants of gently undulating surfaces as they existed when formed by normal stream action at some previous time, but that such original surfaces would lie well above the present ridge tops.

Physiographic history of the Secesh district

The above discussion of the regional physiographic history of central Idaho is based upon the writer's own observations and upon the published reports of many other workers in this general region. C. P. Ross has investigated extensive areas in north-central Idaho and studies of individual districts have been made by A. L. Anderson, P. J. Shenon, J. C. Reed, and the writer, in cooperation between the U. S. Geological Survey and the Idaho Bureau of Mines and Geology, and cover the Elk City, Orogrande, Buffalo Hump, Tenmile, Florence, Warren, Dixie, Thunder Mountain, and Edwardsburg districts of Idaho and Valley counties, and other districts in surrounding areas. In most of these detailed studies, however, the principal attention of the writers has been given to the bedrock geology and the lode deposits, although all added to our understanding of the development of the land forms. During the last two years, the writer's attention has been focused mainly upon the physiographic history and its influence upon the formation of the placer deposits of the region. In no one district can all the pertinent facts that bear upon that history be observed, and the regional problem can best be approached by detailed studies of separate districts and by combining those facts into a single broad picture that applies to the region as a whole.

The Idaho batholith is so homogeneous in composition and texture, and so lacking in recognizable structural features, that its rocks give few clues to the structural deformations that it may have undergone. The pre-batholith metamorphic rocks also fail to give recognizable evidence of late mountain movements. The best evidence of the late deformations that have influenced the development of the present land forms is to be obtained from those relatively young rocks - the Columbia River basalts and the Tertiary sedimentary beds - that were horizontal, or nearly so, when they were laid down. These rocks reflect later movements by their attitude and are most useful in determining the positions of faults and folds, and the direction and degree of crustal deformation. Unfortunately, the lavas do not now, and probably never did, reach as far eastward as the Secesh district. There are, however, certain areas of Tertiary beds that have been uncovered by placer mining excavations, and these, together with the positions of hot springs, give certain or presumptive evidence of the locations of faults, as does also the alignment of certain valley troughs. These facts, taken together with evidence obtained in adjacent areas, make it possible to outline the physiographic history of the Secesh district, with some confidence, as follows:

In early Tertiary time, this district, as well as the surrounding region in central Idaho, was subjected to a long period of erosion and was reduced to a mature surface that has often been referred to as a peneplain, although there is evidence that it always had considerable local relief and that a number of prom-

24

inences stood above it as monadnocks. Part of this surface in the Clearwater
Mountains was later elevated about 1,000 feet and a lower erosion surface developed over considerable portions of it, although that lower surface has not yet been
recognized south of the Salmon River. The Clearwater and Salmon mountains, however, still stood well above the level of the Snake River Basin to the westward.
In Miocene times, great volumes of lava were poured out in the Snake River Basin,
and as they increased in thickness lapped up against the mountains to the
eastward and covered parts of the older erosion surfaces along their western age,
although they probably never extended as far east as the Secesh district. At
some time after the extrusion of the lavas had ceased, perhaps in early Pliocene
time, there was a general uplift of the mountain mass with block faulting and
warping along the western edge of the mountains and probably extending also far
back into them. Some of these faults had displacements of 1,000 feet or more;
most of them trended north or northwest and along most of them the eastern sides
of the block fault were elevated in relation to the western sides so that the
scarps faced east or northeast. There are other places where the presence of
faults has not been proved but where elongated depressions, possibly due to warping also have a north or northwest trend. In the Secesh district, one fault line
follows the valley of Lake Creek to its mouth, and thence extends southeastward
across Ruby Creek to Willow Basket Creek, which it follows to Secesh River. The
course of the Secesh River below the mouth of Willow Basket Creek probably follows
a continuation of this same structural depression. This fault line is marked by
the Burgdorf hot spring and by two other less conspicuous hot springs, one a
quarter of a mile south of Burgdorf and one in Secesh Valley. Still another hot
spring that is probably along the same fault is reported on Secesh River below
Willow Basket Creek.

The basin of Secesh River from Flat Creek to the lower end of the meadows
is definitely known to lie along a fault line, for there are Tertiary beds in the
valley that dip steeply to the southwest, to be cut off in that direction by a
fault that throws the faulted edge of the Tertiary beds against the granitic rocks
A branch of this fault runs northward up Grouse Creek where, likewise, Tertiary
beds dip westward toward it. Just outside the Secesh district, the Warren basin
to the eastward is structural, and French Creek and Elkhorn Creek to the westward
are also believed to follow faulted valleys.

It is thus evident that the drainage pattern in this part of the Salmon
Mountains has been greatly influenced by post-Miocene faulting and warping, and
that many of the present valleys are known to follow structural basins formed
during that period. In some places, tilted lava beds or Tertiary sedimentary deposits prove that relationship beyond argument. Elsewhere the presence of hot
springs strongly suggests the presence of faults, and in still other places de-
pressed rock basins in which there is no drainage outlet give strong evidence of
either faulting or warping.

In the time since the mountain-building processes of regional uplift, faulting, and warping were completed, the shaping of the land forms has been due to
the more normal processes of erosion by streams, glacial ice, and creep. A
major event in this region has been the incision of the Salmon River canyon to a
maximum depth of some 6,000 feet. The erosion of this canyon is believed to have
been begun with the first of the post-Miocene movements of uplift and to have
been continued uninterruptedly ever since. As it happens, the Secesh district
lies in the very headwaters of one of the longer tributaries of the Salmon, and
the rejuvenation of the tributary streams, owing to the lowering of the Salmon
River canyon, has by now barely reached to the edge of the district here described. When the head of the canyon works upstream a little farther, it will tap the
Secesh Meadows, remove the valley fill there, and gradually rejuvenate all the
tributary streams.

The Salmon River Mountains had been elevated to about their present height, and the canyons of the Salmon River and its tributaries had entrenched themselves almost to their present depth by the beginning of Pleistocene times. The Pleistocene epoch witnessed a series of periods of colder climate with glacial advances, alternating with periods of more moderate climate and corresponding ice shrinkage or disappearance. The morainal deposits of two of these stages of glaciation have been definitely recognized in the Secesh Basin, and there are extensive terrace deposits of gravel that are believed to have been deposited during a third glacial stage, intermediate in age between the other two. During the oldest recognized stage of glaciation, tributary glaciers from Marshall Mountain and from the east face of the Bear Pete ridge pushed down into the valley of Lake Creek, there joining to form an ice tongue several hundred feet thick which extended southward to the mouth of that valley. The moraines of this early stage of glaciation are still recognizable at many places along the valley walls. At the same time, another glacier, which originated in the headward tributaries of Secesh River, flowed northeastward as far as the mouth of Ruby Creek and may possibly have once joined the southward flowing Lake Creek glacier. Its moraines appear as a veneer over the low ridge that separates lower Ruby Creek from the Secesh Valley (Pl. IX). A third glacier of this stage headed on War Eagle Mountain and the ridge between Grouse and Lake creeks and pushed down Grouse Creek Valley and down Secesh River to within a mile of the foot of Secesh Meadows. The moraines left by these early Pleistocene glaciers are of special economic importance in this district for they have been extensively mined for placer gold.

An intermediate stage of glaciation in this area has been recognized only by the high-terrace gravels along Grouse Creek and Secesh River (Pl. XIII). The glaciers during this stage were probably smaller than those of the last or Wisconsin stage, and the moraines were probably destroyed or modified by the over-riding Wisconsin ice. The high-terrace gravels also contain some placer gold and have been exploited in places, particularly at the Thorp diggings. The presence in these gravels of abundant granitic boulders that are still quite hard and firm indicates that they are much younger than the oldest moraines, and, although they are certainly older than the youngest Wisconsin glaciation, they are probably closer in age to the Wisconsin than to the earliest Pleistocene stage that has been recognized.

The last, or Wisconsin, stage of glaciation left moraines that still retain their morainal topography, little altered by post-glacial erosion. The Wisconsin glacier that headed on Marshall Mountain moved down Upper Lake Creek to a point below Marshall Lake, but did not reach to the Burgdorf-French Creek road. From the west a half dozen cirques on the Bear Pete ridge sent glaciers down to the edge of the Lake Creek Valley, but none of them was vigorous enough to flow southward down that valley. In the Upper Secesh Basin, glaciers moved northeastward to about the mouth of Lake Creek and left moraines on the Secesh-Ruby Creek ridge that are now being mined for their placer gold. Upper Grouse Creek also nourished a Wisconsin glacier that pushed southward to within a little more than a mile of Secesh River. Several small mining ventures have been carried on within the area covered by that glacier, but most of the work on Grouse Creek has been below the limit reached by the Wisconsin ice.

The outwash gravels from the Wisconsin glaciers have been little dissected in post-glacial time, and in many places blend imperceptably with the present stream gravels. These gravels and the present stream gravels contain some placer gold, but they have been little exploited for their gold content is low and in places the abundance of large boulders has so far made them unattractive to the miners.

ECONOMIC GEOLOGY

The first discovery of placer gold in central Idaho at Pierce, in the Clearwater Basin in 1860, was followed by a period of prospecting that in 1862 resulted in the finding of rich stream placers in the Warren district, and soon afterward of the placer deposits in the adjacent Secesh Basin. This was during the Civil War period, and accounts for such names as Secesh and Dixie, apparently given by sympathizers with the Confederacy. The Secesh Basin has generally been included in the Warren district, both in discussions of the history of the camp and in tables of production. Although the placers of the Secesh Basin have been actively exploited at various times from their discovery to the present, they were never as rich as those near Warren, and early references in the literature make little or no separate mention of the Secesh placer mines. As a consequence, much of the early history of the camp, including the dates of discovery and time of active exploitation of the various claims, has been lost. Furthermore, such unsatisfactory records of early placer production as are available from this region include the output of the Secesh placers with that of the Warren district, and only a rough estimate can be made of the gold output to date of this basin. Figures of production since 1901 have been kept by the U. S. Bureau of Mines, and their total for that period, to which has been added an estimate for 1938, gives a total placer output of $152,600.00. Compilation and evaluation of estimates made by a number of men who have long been familiar with activities in this area place the value of placer gold output in the years 1862 to 1900 at $326,000. This figure is, of course, only approximate, but it appears that Secesh Basin has to the present time produced placer gold valued between $450,000 and $500,000.

In central Idaho, placer mining by hydraulic methods is generally hampered by the scarcity of water in middle and late summer. The heavy winter snowfall gives abundant water almost everywhere in the spring and early summer when the snow is melting, but the precipitation in the summer months is scanty and comes only as occasional thunder storms. As the small miner often has limited resources long ditches are out of the question and he usually has to rely on some local small stream to furnish water under head for sluicing. As a result, many placer mines can be operated only from early June to middle or late July, and the short season greatly increases the cost of moving ground. Gold dredges and draglines, by recirculating their water through pumps, can operate on a much more restricted water supply, and so take advantage of a much longer mining season.

Gold placer deposits result from the erosion of gold-bearing lodes, the comminution of the lode material, the removal by streams of much of the barren gangue material, and the concentration of the heavy and chemically inert metal in the stream beds. Obviously, it is necessary, in order that workable placer deposits exist in any valley, that there be gold lodes in present or ancient drainage basin and that those lodes be exposed to the processes of erosion. In the Secesh district, these conditions are present in two separate areas. The Marshall Lake district, centering around Marshall Mountain, is an important gold lode camp in which there are not only several mines that have produced lode gold, but many promising prospects. Lake Creek and Grouse Creek rise in that mineralized area. The other area, in the headward basin of Secesh River, has not been as thoroughly prospected as the Marshall Lake district, but some lodes containing gold and others containing mercury have been found there, and the presence of placer gold in the glacial deposits indicates that there are other lodes containing free gold that have not yet been found. The heavy concentrates from the placer mines bear out the conclusion reached from the physiographic evidence, that the gold found on Lake Creek and on Grouse Creek comes from a different locality than that on Ruby Creek and in the upper Secesh Basin. The sluice box concentrates on Lake Creek show abundant crystals of corundum and little or no monazite or cinnabar.

27.

By contrast, the placers of the Ruby Creek area carry abundant monazite and cinnabar, and little or no corundum. Below the mouth of Lake Creek, one would naturally expect a mixture of materials derived from these two sources.

The placer deposits of Secesh Basin present many unusual features. In contrast to most placer camps, where the gravels of the present streams contain most of the gold, the present stream gravels in this camp have been worked very little, and it is known that in most of them the gold content is low. Along the lower side slopes of many valleys, the attention of the early prospectors was attracted to deposits in which almost all of the boulders were quartzite, although the prevailing bedrock in this basin consists of granitic rocks. Very naturally, they attributed the distribution of this quartzitic material to old stream channels that headed in some unlocated area in which quartzite bedrock predominated. Prospecting showed that this quartzitic material generally contained some placer gold, and in places was rich enough to mine. The common procedure was to dig a ditch to the nearest adequate stream to bring water under head to the deposit, and by hydraulic methods, using steel pipe, canvas hose, or ground sluicing, to sluice the fine material and to pile the boulders to one side. In those early diggings, best exemplified on the ridge between lower Ruby Creek and Secesh River, on Lake Creek a few miles above Burgdorf, and at various places on either side of Secesh Meadows, from 4 to 10 feet or more of material were moved. In some places, the material was moved to granitic bedrock and in others Tertiary sedimentary bedrock was exposed, but in many pits the floor was in much the same sort of material as that which had been mined. When granitic bedrock was exposed, it was commonly found to be so deeply disintegrated that it could be mined like sand. In most of the early mining, few pebbles or boulders larger than 5 or 6 inches in diameter were put through the boxes, and most of the boulders were piled in long rows in the cut. The placer pits were excavated at right angles to the trend of the valley and were narrowest at their upper ends near the ditch line, and fanned out below. Even a casual inspection of these old placer pits shows that an unusual set of conditions prevails here. Many of the placer deposits are unlike ordinary stream gravels in topographic position, in lack of assortment and in the presence of abundant quartzite boulders and the scarcity of boulders of the prevailing granitic country rock. Instead of lying along the valley bottoms like ordinary valley fill, or as terraces representing older valley floor deposits, some of these placer deposits lie on the steeply sloping valley walls, extending as much as 300 or 400 feet above the streams. Their surface expression is much like that of any other mountain slope in this region with little except the numerous quartzite boulders to suggest the presence of material transported from a distance. Fortunately, a considerable number of fresh exposures of these deposits are found in old placer cuts and reveal their character and origin, disclosing the fact that placer deposits of several types have been mined.

Placers in early Pleistocene moraines

Some description has already been given (pp. 13-15) of moraines of early Pleistocene age that have been recognized at this and many other places in the mountains of central Idaho. In the Secesh Basin, these ancient moraines were deposited by glaciers that were considerably longer and thicker than the last or Wisconsin glaciers, and are now found in places on the valley walls beyond the Wisconsin moraines and to a height of as much as 400 feet above the valley floors. These moraines are characterized by the presence of quartzite boulders on their surfaces, by the absence of granitic boulders, and by the absence of recognizable glacial topography on their surface. These old moraines contain some placer gold and in places enough to have encouraged mining. Excavations in them show that the materials are deeply oxidized, and that the granitic boulders which originally predominated greatly over other kinds have completely disintegrated to

28.

depths of as much as 30 feet or more below the surface, and in mining break down
and pass through the sluice boxes as sand. The surprising fact that these morainal deposits contain placer gold in commercial amounts is thought to result from
the movement of glaciers down valleys in which stream placers had already accumulated; the gold first concentrated by streams was incorporated in the glacial
moraines. Some secondary concentration of this gold in the moraines took place
in a surficial layer about 6 feet thick as a result of surface creep, and the removal of some waste material by hillside wash, wind, and other agencies. Mining
in these old moraines in which the gold content is moderate was possible in the
early days of the camp only because of the steep gradients on the hillsides which
allowed easy disposal of tailings and gave large duty from the available water,
and also because at least 90 per cent of the large boulders were so rotten that
they disintegrated completely and could be sluiced through the boxes. Apparently
also, there was enough secondary concentration in the surficial creep zone to
make the mining of it profitable in places where the mining in undisturbed moraines would not pay and the practice of the early miners was to sluice little more
than the surface zone. In places, the disintegrated material from early morainal
deposits has crept down hill for some distance over the top of younger Wisconsin
moraines, and locally this overlying material has been mined down to the underlying fresh moraine which itself, though also containing some placer gold, is too
bouldery to be handled profitably. In this district, the placer mining operations
that worked these old morainal deposits, and the surface creep material derived
from them, include most of the old workings on the ridge between Lower Ruby Creek
and Secesh River, part of the material worked on the Golden Rule ground along
Grouse Creek, and the old Gayhart Burns diggings on the east side of Secesh
Meadows.

Placers in early Pleistocene terrace gravels

Another type of deposit that has in past years been mined rather extensively
in this district includes high-terrace gravels that now stand 30 feet or more
above the present streams. These high terraces apparently include materials that
range in age through at least two pre-Wisconsin stages of glaciation, and they
probably represent the outwash gravels laid down during Pleistocene stages of
glaciation that antedate the Wisconsin stage. In some of these high-terrace
gravels the granitic boulders are completely disintegrated to depths of as much
as 30 feet, below which a few are still fairly solid. The quartzites and some
other resistant types of rock are still firm, and have been piled in the old
placer cuts. Locally, at their upper edges, these old terrace gravels grade imperceptibly into old moraines and appear to represent outwash from the same glaciers that deposited the moraines. Placer mines in these oldest high-terrace
gravels include the old workings of the Threemile mine just west of the road at
a point 2.4 miles north of Burgdorf, another old pit that crosses the road 4
miles north of Burgdorf, and parts of the old Golden Rule placer workings on
Grouse Creek.

Placers in middle Pleistocene terrace gravels

A third type of placer deposit in this district includes high terraces that
are intermediate in age between the oldest recognized moraines and the youngest
Wisconsin moraines. These terraces are best developed on lower Grouse Creek, and
on the southwest side of the Secesh Valley between Flat Creek and Long Gulch. In
these terraces, oxidation is less advanced than in the oldest terrace gravels,
and many of the granitic boulders are wholly or partly disintegrated, although
many are still firm. These terraces are believed to represent outwash from glaciers that were of pre-Wisconsin age, but that were younger than those that left
the oldest recognized moraines. They contain some placer gold, but too little to
have been exploited profitably. They were most extensively mined at the Thorp
property 1/2 mile south of the mouth of Grouse Creek.

Placers in Wisconsin moraines

A fourth type of placer ground has here been exploited only during the past two years (1937-38) on the Davis Mining Company ground at a point 3,600 feet, south-southeast of the mouth of Lake Creek. Here mining has been conducted recently on ground that had already been worked, as the surficial layer of creep material from old moraines that overlay the younger Wisconsin morain had been sluiced off many years ago. Recent developments have proved that the Wisconsin morain itself contains considerable placer gold and that beneath it and above the granitic bedrock there is in places a layer of pre-Wisconsin, oxidized, quartzite gravel that is relatively rich in gold. Here, in a single section, there are three successive layers of gold-bearing material of different ages.

Placers in present stream gravels

A fifth type of placer deposit that has been little mined in this district includes the gravels of the present streams. Mining of the present stream gravels has been confined to some headward tributaries of Grouse Creek, and particularly to the valley of Ruby Creek where small-scale pick-and-shovel operations have been carried on at various times in the canyon-like lower portion of the valley, and where a small dragline scraper was operated in 1937. There are in the district, nevertheless, large areas of present stream gravels in the meadows that are known to contain some gold, and some of these meadows will doubtless eventually be mined by means of low-cost, mechanical methods.

Placers of composite type

In the preceding pages, the different types of placer deposits in this district have been classified according to their origin and relative age, for effective mining depends upon an understanding of the way in which a placer was formed. Yet miners and prospectors are necessarily guided in exploration and development by the gold content of the ground, as revealed by constant testing with the gold pan, and, on single claims, minable ground has been found that falls in two or more of the types mentioned. It is only by a careful study of the ground while exposures are fresh that the history of the deposits can be confidently deciphered. Much needed information concerning the character of the material, the relations of one type of deposit to another, and the gold content of the various materials has been lost concerning those properties that were worked so many years ago that their exposures are now slumped and obscured.

Description of placer mines

Placer mining has been carried on continuously in the Secesh Basin for the past 75 years, though with varying degrees of vigor. In the early days of mining, no systematic records of locality, personnel, or production were kept, and the records of mining activity in the years before 1901 are vague and incomplete. The most active period of placer mining was in the years from 1880 to 1912, as is evident from the extent of the old pits that were excavated during that period, and the bulk of the gold that has come from this camp was recovered during those years. An attempt was made to visit as many as possible of the old residents in the district, and to compile from their recollections a chronology of mining and of production, but their accounts were often at variance, and their knowledge of production was vague. The historical information given in the descriptions of the various properties is believed to be as reliable as can now be obtained. In 1938, two hydraulic mines were active in this basin and perhaps a half dozen smaller pick-and-shovel operations were in progress, but it is doubtful whether more than 15 men in all were engaged in placer mining at any one time. In the succeeding pages the gold placer deposits will be described in the following order:

30.

(1) Lake Creek Basin; (2) Secesh-Ruby Creek area; (3) Grouse Creek Basin; (4) Secesh Meadows.

Lake Creek Basin

The headward tributaries of Lake Creek include Corduroy Creek, West Fork of Lake Creek, and the head of Lake Creek itself. On all of these streams there are stream gravels and some terrace remnants that have been inadequately prospected, but are known to contain some placer gold. A number of prospect pits have been sunk at various places, and one large pit reflects an attempt at hydraulic mining on a bench of Upper Lake Creek, but apparently the amount of gold found was insufficient to encourage further developments. These tests were, however, made at a time when the district was much less accessible than it is now and when gold was valued at $20.67 an ounce. In view of the greatly improved roads of today, the increased price of gold and the development of mechanical methods of handling ground more cheaply, it is possible that some of this ground can now be mined at a profit, and thorough prospecting is warranted.

From the confluence of West Fork of Lake Creek with Lake Creek to a point a short distance below the mouth of Three Mile Creek, Lake Creek flows through a stream flat from 150 to 450, or more, feet in width, and is flanked on the east by a broad gravel bench that grades imperceptibly into the hillside slope above it. The gravels of this stream flat are undoubtedly gold-bearing, although no evidences of prospecting were seen. However, they certainly must have been tested in the early days of the camp and the fact that no attempt was made to mine them implies that the gold content is low. They are suitable for dredging, if sufficient gold is present.

Lake Creek Company

The bench gravels to the east of this stream flat have in the past been mined somewhat extensively at two localities, one of which is crossed by the road at a point 4 miles north of Burgdorf, and the other at a point 2.4 miles north of Burgdorf and 500 feet below the road. It is probably to the first of these operations that Lindgren [1] referred when he stated that the Lake Creek Company began operations in 1897, intending to wash the benches by hydraulic method. The gold content of the gravel was estimated as $.15 per cubic yard. Plate VII is a map of the excavation on this property which has been inactive for many years. The area of the pits is about 80,000 square yards, and, if an average depth of moved material is assumed to be 21 feet, the excavations aggregate 560,000 cubic yards. This material, after being put through the sluice boxes, was dumped on the stream flat below the edge of the terrace, and the portion that has not been swept away by Lake Creek will have to be rehandled if that flat is ever mined. The terrace edge stands about 30 feet above the stream and extends from the east edge of the stream flat eastward to the road, above which the hill slope rises more steeply. As is shown on the accompanying sketch, the main cut below the road was excavated to, and some distance into, the granitic bedrock, which is here so deeply decomposed that it can be easily sluiced. The section as exposed in the sides of the cut shows from 4 to 12 feet of ancient moraine material, well oxidized, in which the granitic boulders are completely disintegrated; the only boulders left intact are of quartzite, schist, or other resistant rocks. This material lies on decayed, granitic bedrock into which the mine excavation penetrates locally to a depth of 20 feet without reaching undecomposed material. The surface layer of old, disintegrated, moraine material has been laminated to a depth of 5 or 6 feet of creep.

In the pits above the road, the disintegrated, granitic bedrock is exposed at only one place, and has there been excavated for road material. Elsewhere the

1/Lindgren, W., Twentieth Ann. Rept., U.S. Geol. Survey, Pt. III, p. 242, 1900.

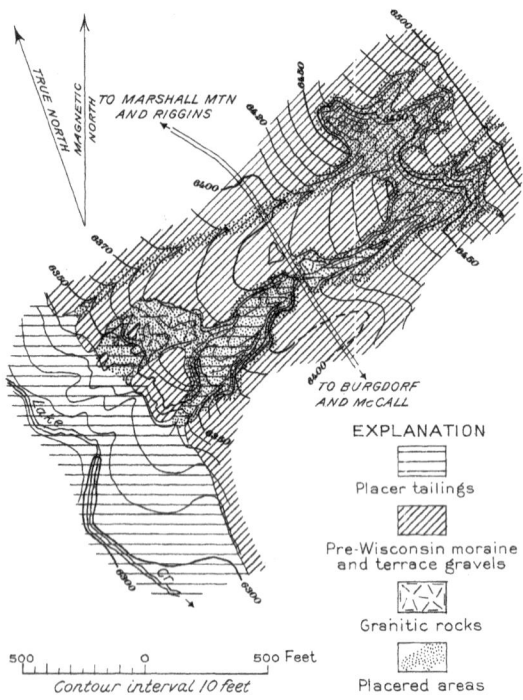

EXPLANATION

Placer tailings

Pre-Wisconsin moraine
and terrace gravels

Granitic rocks

Placered areas

TO MARSHALL MTN
AND RIGGINS

TO BURGDORF
AND McCALL

TRUE NORTH

MAGNETIC NORTH

500   0   500 Feet

Contour interval 10 feet

LAKE CREEK PLACER COMPANY WORKINGS, LAKE CREEK,
IDAHO COUNTY, IDAHO.

pits have failed to reach bedrock, although the maximum depth to which mining was carried was about 50 feet. The walls and floor of the pit are composed for the most part of old, well oxidized, morainal materials of buff color to a depth of 20 feet, but below that are blue-gray, tough, and unoxidized. In the oxidized portion, the granitic boulders have rotted completely and disintegrated to sands under the hydraulic stream. The firm, quartzite boulders, many of which are 3 feet or more in diameter, were stacked to one side in the cut. About 95 per cent of the boulders so stacked are of quartzite, but there are also a small number of firm, granitic boulders that are believed to have come from the deeper, unoxidized areas in the pit. An examination of the hillside above the upper edge of the pits showed that quartzite boulders are present to an altitude of 6,700 feet where there is a bench-like break in the slope that probably represents the top of an old lateral moraine. The location of this moraine indicates that the early Pleistocene glacier that deposited the morainal material here reached a thickness of 370 feet. The 30-foot difference in altitude between the low edge of the terrace and the present flat of Lake Creek probably represents the amount of lowering of its valley floor by Lake Creek since this early Pleistocene stage of glaciation.

No reliable information could be obtained on the amount of gold that was recovered in this enterprise. Evidently, it was not profitable under the conditions then prevailing, or it would have been continued. Apparently, there is a great volume of this old morainal material in this vicinity, and it is so situated that it could be cheaply handled by mechanical means or by a modern hydraulic plant. The property deserves thorough prospecting.

Three Mile placer

The old workings of the Three Mile placer lie on a broad, high terrace near the mouth of Three Mile Creek, and just west of the road at a point 2.4 miles north of Burgdorf. The terrace is here about 1,500 feet wide, and its lower edge is about 30 feet above the level of Lake Creek. The location, size, and shape of the old pits, and their relations to the road and to Lake Creek, are shown in Plate VIII. Hydraulic mining is said to have been carried on here at different times from 1870 to 1917, and in 1938 several individuals who had lays on the ground were working in a small way by hand methods.

The larger of the two pits has an area of about 22,000 square yards, and averages about 12 feet deep, indicating that about 66,000 cubic yards of material were mined from it. The smaller pit, with an area of about 8,000 square yards, is shallower and has a capacity of about 16,000 cubic yards. The walls of the pits show that the material mined consisted of an aggregate of coarse gravel and boulders as much as 3 feet in diameter, and included mainly granitic, gneissic, and schistose boulders with a smaller proportion of quartzite boulders. The granitic boulders, and most of those of gneiss and schist, are so decomposed that they disintegrate on mining, and the dumps consist almost exclusively of quartzite. The material is weathered to a buff color, and many of the quartzite boulders are coated with manganese oxide. Sluice box concentrates contain abundant corundum in well formed crystals, smaller amounts of garnet, hematite, ilmanite, limonite, almost no magnetite, and only traces of zircon and monazite. No bedrock was seen in these cuts, and the bottoms of the excavations were in a fine, micaceous sand interlayered with blue clay, which may be in part fines from the placer mining operations. It is said that a pit sunk to a depth of several feet in the floor of the larger cut failed to reach bedrock. This mine is on a southward continuation of the same terrace as that upon which the lower part of the Lake Creek placer is situated. The deep weathering of the material and the disintegration of the contained granitic boulders indicate an early Pleistocene age for the deposit, which may be in part an old moraine and in part moraine modified

TO MARSHALL MTN
AND RIGGINS

TO BURGDORF
AND McCALL

CAMP 6970

EXPLANATION

Placer tailings

Pre-Wisconsin moraine
and terrace gravels

Placered areas

TRUE NORTH

MAGNETIC NORTH

Lake Creek

500   0   500 Feet

Contour interval 10 feet

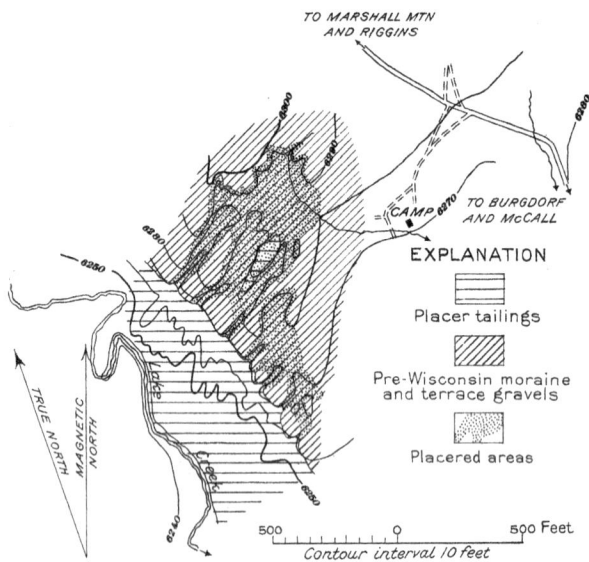

THREE MILE PLACER WORKINGS, LAKE CREEK,
IDAHO COUNTY, IDAHO.

by streams and intermingled with glacial outwash gravel. The absence of magnetite in the concentrates may be due to its weathering to limonite during the long period since the deposit was formed.

No information was obtained of the average gold content of this deposit, although it was sufficient to encourage a renewal of operations at several periods since the first mining was done. A large volume of similar material is available on the bench, both to the north and south of this old mine, and is well adapted to development by dragline equipment. Thorough prospecting should be done to prove whether or not the gold content is adequate to justify the installation of mechanical equipment.

Secesh-Ruby Creek area

Davis Mining Company

The Davis Mining Company owns 25 claims on the ridge that lies between Secesh River and Lower Ruby Creek, and opposite the mouth of Lake Creek. The property is now under option to Don S. Numbers of McCall. It has been mined at various times, and at different places, since the Civil War, although the greatest activity was before 1900. Plate IX, a geologic and topographic map of the vicinity, shows the location of the various mining operations that will be described here. Although most of the old cuts are weathered and slumped, there are still excellent exposures that reflect the unusual conditions that prevail in this ground and throw much light on the origin of the placer deposits throughout this district.

The ridge on which this ground lies here rises to an altitude of 6,330 feet, which is 320 feet higher than Secesh River at the mouth of Lake Creek. It is composed of granitic rocks, over much of which there is a veneer of glacial morainal material that ranges in thickness from a thin layer to 50 feet or more. This glacial moraine is of two distinct ages, the older being of early Pleistocene age and the younger of late Wisconsin age. Apparently, this area marks about the lower limit to which the glaciers of both these states reached as the glacial deposits are of terminal moraine type. The older glacier was a hundred feet or so thicker than the Wisconsin glacier and pushed somewhat farther down valley.

The characteristic and somewhat puzzling feature of the older glacial deposits, which alone were exploited in the early days, is the absence on their surface of any boulders except quartzites, although the basin from which they were derived is composed mainly of granitic rocks. The boulders stacked in most of the old mining pits are composed almost exclusively of quartzite. An examination of freshly exposed faces in the pits gives a complete explanation of this condition.(p. 29). Plates IV and V illustrate the conditions of this locality. It was only because most of the big boulders were so decomposed that the material could be mined at a profit. Another unusual condition in this placer ground is the concentration of gold by surface creep in a surficial layer about 6 feet thick. In this creep zone, the disintegrated granitic boulders have been drawn out by the down-hill movement to form a laminated zone, and apparently enough of the fine, comminuted material in this zone has been removed by sheet and rill wash to give a greater concentration of gold in the surficial material than was present in the undisturbed moraine. In many pits, this creep zone was removed in mining and the undisturbed portion of the old, disintegrated moraine was left behind.

Here also a creep zone layer of the older, disintegrated moraine has moved down over the surface of the younger, fresh Wisconsin moraine. This material

33.

was found to be minable down to the surface of the younger moraine where the hard granitic boulders were so abundant that mining was not profitable.

The astonishing fact about all these deposits is that morainal material, in which there has been little concentration of gold by running water, should be rich enough in gold to be minable by hydraulic methods. This condition can best be explained by assuming that there were pre-glacial placer concentrations of considerable richness in this basin, and that those old stream placers were picked up by the glaciers and incorporated in the moraines. This assumption is supported by the discovery in places of patches of rich placer gravel that lie on bedrock beneath the Wisconsin moraines. At only one locality has bedrock been uncovered beneath the Wisconsin moraine, and there the underlying stream gravel is patchy but is deeply weathered, is composed almost exclusively of quartzite gravel and is locally very rich. A number of pans of material taken from the surface of the bedrock yielded an average of 10 cents in gold to the pan.

On the northwest slope of the ridge, the upper edge of the Wisconsin moraine can be traced with accuracy through the old placer pits, although elsewhere it is obscured by old morainal detritus that has crept down over it from the hillside above. It nowhere extends up the ridge slope more than 100 feet above its base. The older moraine by contrast extends to the top of the ridge and well down its eastern slope, where it has been so attacked by erosion on the steep Ruby Creek slope that its original outlines can no longer be accurately recognized. Considerable areas of the Ruby Creek slope have now only scattered quartzite boulders lying on granitic bedrock, and many of these boulders may have been carried to their present resting places by hillside creep.

Sluice box concentrates were collected from three places in this vicinity that represent somewhat different conditions. The minerals identified are as follows:

	Pre-Wisconsin Moraine	Wisconsin moraine and underlying pre-Wisconsin gravel	Gravels of Ruby Creek
Quartz		x	x
Magnetite	x	x	
Ilmenite	x	x	x
Garnet	x	x	x
Zircon	x	x	x
Monazite	x	x	x
Cinnabar	x	x	x
Rutile	x		
Feldspar	x	x	x
Augite	x		
Hornblende	x		
Andalusite	x		
Muscovite	Trace		
Gold	x	x	x

It should be noted that the sample from "Wisconsin moraine and underlying pre-Wisconsin gravel" was on ground from which creep material from pre-Wisconsin moraine had already been mined, and may, therefore, have derived some heavy minerals from it. The gravel of Ruby Creek also probably contains some material derived from both Wisconsin and pre-Wisconsin materials. These three sets of concentrates are remarkably similar in their content of heavy minerals, particularly as they contain monazite and cinnabar in considerable quantities. The failure of some of the samples to contain such relatively light minerals as quartz, feldspar, augite, hornblende, and muscovite, is probably due to the more thorough panning of those concentrates before they were tested. There can be little doubt that both the Wisconsin and the pre-Wisconsin morainal material came from the same source; that is, the headward basin of Secesh River.

So far, the only valuable material recovered from this ground has been placer gold, but the miners have been interested in the possibility of marketing the monazite for its cerium and thorium content, and of retorting the heavy concentrates to recover the quicksilver from the cinnabar. So far, no market has been found for the monazite, and it is doubtful whether enough mercury could be recovered to justify the cost of retorting the concentrates.

Except for desultory, small-scale mining by hand methods by two or three individuals, the only active mining on this property in 1938 was the hydraulic operation of E. F. Stratton, Lee Kessler, and associates, lessees, on the west slope of the ridge 4,000 feet south-southeast of the mouth of Lake Creek. About ten men were employed in March and April in opening up an old ditch that brings water 4-1/2 miles from Ruby Creek. After mining began on May 1, from three to four men were employed in mining. Water for piping is delivered with a head of 154 feet at the cut through steel pipe, and a 5-inch nozzle is used on the giant. A string of 320 feet of 24-inch sluice boxes, set on a grade of 6 inches in 12 feet, and lined partly with steel-faced ripples and partly with pole riffles, was used for washing the gold. Boulders less than one foot in diameter were put through the boxes. Those larger than that were stacked to one side by hand, or if too large to be so handled were broken by sledge or with powder. This ground had already been mined over once, 40 or 50 years ago, by the Larson brothers, who constructed the ditch that is still in use, but, in their mining operations, they removed only the surficial 6 to 15 feet of material. In the lower portions of their pits, they mined little more than the surface creep zone of detritus that had moved from the older, decomposed morainal material of the upper slopes down over the surface of the younger Wisconsin moraine. Wherever their excavations penetrated to the Wisconsin moraine, they encountered abundant large, sound, granitic boulders, and these boulders made mining too expensive to be profitable. The Larson pit at this place nowhere reached bedrock.

In 1936, a layman sluiced an excavation in one of the old Larson pits and penetrated through the Wisconsin moraine to the granite bedrock, upon which he found patches of rich, pre-Wisconsin quartzite gravel. The Stratton workings adjoin that excavation. In 1938, the Stratton pit showed a face 40 feet high, the upper 38 feet of which was Wisconsin moraine that contained abundant hard, granitic boulders as much as 5 feet in diameter (Pl. X). Beneath this moraine on the south and west sides of the pit, there was a layer as much as 4 feet thick of well-washed quartzite gravel that everywhere showed a high content of gold on panning. The overlying Wisconsin moraine also contained considerable gold, although its distribution was erratic. The underlying granitic bedrock was somewhat decomposed, particularly beneath the quartzite gravel layer, so that it could be cut with a pick. The operation in 1938 was only moderately successful for the equipment used was old and unsatisfactory, the cost of disposing of the abundant boulders was high, and the water supply was so inadequate during middle and late summer that it could be piped for only a few hours a day. It is evident that the cut, if extended up the slope another 250 feet, will pass through the Wisconsin moraine and into the weathered older moraine that caps the ridge. In the old moraine, the granite boulders are so thoroughly weathered that they will disintegrate under the hydraulic jet and only the resistant quartzite boulders will need to be handled. This will greatly decrease the unit cost of moving the ground, but it has not been determined whether or not the gold content of the old morainal material is sufficient to justify mining it.

Plate X. Hydraulic placer mine in Wisconsin morainal material. Davis Mining Co., Secesh River.

That it does contain considerable gold is evinced by the extensive pits mined in it in the past. The gold from this property assays about .840 fine and is bright and fairly heavy. Nuggets worth $1.00 are rare, although there is little very finely comminuted gold. Early in September, it was reported that 25,000 cubic yards of material, containing $.18 per cubic yard, had been mined.

One or more of the conditions already described at Stratton's workings can be found at all of the old pits on this property. The extensive old pit a few hundred feet south of the Stratton pit, worked by the Larson brothers, was for the most part in the older morainal material, lying on disintegrated granitic bedrock, although there, too, detritus that had crept down over the Wisconsin moraine was also mined. In the large pit at the head of West Ruby Meadows, the old moraine was thin, or altogether lacking, and the surface was covered with as much as 4 feet of soil containing quartzite pebbles which overlay deeply decomposed granitic bedrock. Probably much of this quartzite-bearing soil had spread down the slope by creep from the moraine-covered crest of the ridge. In mining, the surface material and locally as much as 20 feet of the weathered bedrock were sluiced.

The old excavation at the head of Miller Gulch, a tributary of Ruby Creek, is said to have been made by one Miller and associates about 1880. The banks are now much eroded and slumped, but apparently the upper end of the pit was excavated in old moraine material to a depth of at least 20 feet. In the lower portions of the cut, mining extended down to deeply decomposed, granitic, bedrock, and considerable residual arkosic sand was removed. The boulders encountered in mining the upper part of the cut were almost exclusively of quartzite or other resistant rocks, and, when the volume of material mined is considered, the number of boulders that needed to be handled was remarkably small. In the lower end of the pit, many large, residual, granitic boulders were encountered. Apparently, the gold recovered all came originally from the old moraine deposit on the upper slopes of the ridge, and was somewhat concentrated in Miller Gulch by creep and by stream erosion. No information is now available as to the average gold content of the ground mined.

A considerable area was mined in the early days of this camp in the upper basin of Burgdorf Gulch, a tributary that joins Ruby Creek from the west about 3,000 feet above its mouth. There, conditions were much as in Miller Gulch. The head of the cut was excavated into old moraine, which is at least 15 feet thick. Farther down the gulch, the moraine thins out and a sheet of creep material was mined down to disintegrated, granitic, bedrock. The amount of gold recovered is not known.

Plans were under consideration to work this property on a much more extensive scale than has been done in the past. The owner believes that the entire capping of this ridge, including the deposits of both Wisconsin and pre-Wisconsin moraine, can be worked at a profit if a larger and more continuous supply of water is made available, and that such a supply under increased head can be obtained from Ruby Creek by a new ditch and pipe line. Prospect pits near the ridge top are reported to have penetrated as much as 40 feet of old morainal material, all of it gold-bearing. Certainly, the enterprise is attractive in that it has a large body of gold-bearing material with plenty of gradient for sluicing and abundant dumping ground. It is suggested, however, that systematic sampling of the deposit by drilling should be undertaken before a large investment is made in securing a new water supply and in installing new mechanical equipment.

Ruby Creek

For the lower 1-1/2 miles of its course, Ruby Creek flows through a narrow valley in which the stream flat is little wider than the stream itself and the granitic valley walls rise steeply on either side. In this stretch, small-scale mining by hand methods has been carried on from time to time since the early days of the camp, but the amount of placer ground was small and the production unimportant. In 1937, a small, improvised dragline scraper and washing plant was operated by J. T. Jones and sons in a stretch extending for a few hundred feet above the mouth of Miller Gulch. The cut averages about 50 feet in width and the average depth of the ground was 4-1/2 feet. Few large boulders were encountered, and mining was carried down to decomposed granitic bedrock. The sluice box concentrates contained large quantities of monozonite and considerable cinnabar. It is said that about $1,500 in gold was recovered.

Above the Jones workings, the flat of Ruby Creek widens out and expands into a wide flat in West Ruby Meadows. Prospect pits and drill holes in this flat show that it has been tested, but it is reported that the gold content was too low to justify development. The ground is said to be from 6 to 25 feet deep and is overlain by a considerable thickness of barren tailing sands washed down from the hillside placers above.

Grouse Creek

The history of prospecting and mining on Grouse Creek is now difficult to recount as the accounts of various early settlers differ in many details. Apparently, ground was staked on Secesh River and on Grouse Creek during Civil War times by a group of men named Bundle, Brown, and Gayhart Burns. Bundle and Brown concentrated their attention on the ground now known as the Golden Rule placer, while Burns worked the hillside placer a mile north of Long Gulch and just east of Secesh Meadows. Bundle and Brown seem to have worked their ground by hand methods until about 1901 when a large hydraulic plant was installed. In the mean time, several parties, including N. B. Willy, F. Ault, W. Flint, J. Claire, and W. Edwards, mined the stream gravels on some of the headward tributaries of Grouse Creek, including some ground that was quite rich. Detailed information about these operations is now lost. In 1938, some small-scale mining by hand methods was in progress in the gulch northeast of Kelly Meadows in upper Grouse Creek.

Golden Rule placer mine

The Golden Rule placer workings include one large pit and a smaller one on the east valley wall of lower Grouse Creek. It is reported that this ground was staked some time between 1862 and 1870 by Bundle and Brown, and that it was worked by them by hand methods until about 1900 when a company was organized and a large hydraulic plant installed. A ditch about 2 miles long was constructed to bring water under head from Grouse Creek, and a pit 3,800 feet long, from 200 to 1,000 feet wide, and from 6 to 30 feet or more deep, was excavated in the side hill. The company operations were carried on with varying degrees of success until 1911, and in 1913 the property was sold to L. E. Winkler and associates, who have mined in a small way each year since. Winkler's mining, however, has not been on the side hill deposits, but has followed an old gravel-filled channel that lies at the junction of a broad, high terrace with the slope to the east. In the mean time, two men named Copenhaver and Mathias mined a large, shallow cut on the ridge top just south of the Bundle and Brown cut, on ground that is now part of the Golden Rule property.

The total gold production from this property is not known. There is no record of the very considerable amount recovered before 1904, the first year for which accurate figures are available. The U. S. Bureau of Mines statistics show a production of 7066.53 ounces of gold in the period 1904-1937.

The geology of the Golden Rule area (Pl. XI) shows a complex set of factors, all of which have influence upon the development. The productive area was all covered in early Pleistocene time by a glacier that originated in the head of Grouse Creek, pushed down the valley of Grouse Creek to its mouth, and continued down the Secesh Valley almost, if not quite, to Long Gulch. The margin of that glacier had an altitude of at least 6,100 feet along the east side of lower Grouse Creek. Its lateral moraine, deposited along the valley wall, had a varying thickness and contained some placer gold. After the retreat of that early glacier, there followed a long interglacial period during which the old moraine was weathered so that most of the granitic boulders disintegrated, as did the underlying granitic bedrock. Surface creep and stream and sheet erosion carried away much of the old moraine, in places leaving only scattered boulders of resistant quartzite on a surface of decomposed granitic bedrock. Later, in Pleistocene time, there was another, though less extensive, glacial advance, during which a thick deposit of coarse gravel outwash filled Grouse and Secesh valleys to a depth of 30 feet or more, and in places overlay the older moraines. This terrace gravel also contained some placer gold. A last glacial advance, in Wisconsin time, yielded gold-bearing outwash gravels which locally filled pre-existing valleys in the terrace deposits. Post-Wisconsin erosion has produced some further concentration of gold in the present stream flats, although they are not rich enough to have encouraged exploitation.

Cross sections, drawn across the Golden Rule placer pit at different places, show a surprising variety of geologic conditions. The material moved in placer mining included stream gravels not older than the Wisconsin glacial stage, in which the granitic boulders are only slightly weathered; high terrace gravels of pre-Wisconsin age in which many granitic boulders have broken down, though many are still firm; early Pleistocene moraines, which large granitic boulders are thoroughly disintegrated and only quartzite boulders have survived; creep zone areas in which quartzite boulders and fine materials, the remains of old glacial deposits, form a thin layer above decomposed granitic bedrock; and large volumes of arkosic sand that represent decomposed granitic bedrock in place. The bedrock down to which mining was carried includes disintegrated granite, inclined beds of Tertiary sand, shale, and lignite, old morainal materials, and locally hard granitic bedrock from which the disintegrated weathered portion had been removed by erosion before later gravels were deposited over it. Apparently, the early miners were first attracted by the quartzite gravels which they found on the hillsides, and which they proved to be gold-bearing. This quartzite wash was the residue from the older moraines in which the granitic boulders had disintegrated and broken down to sand, and on which the surface layers had crept down the slope. Later mining proved that there was some gold in the undisturbed old moraines, in the upper part of the decomposed bedrock, and also in the high-terrace gravels; all of these were mined. The area of the pit mined before 1913 is about 250,000 square yards, and, if an average depth of 15 feet is assumed, it appears that about 1,250,000 cubic yards of material were sluiced. A cross section through this pit is shown in Plate XI.

Since 1913, L. E. Winkler, with a crew of from one to four men, has mined an old gravel-filled channel that lies on the east edge of a high terrace, and at the base of the valley wall. This channel was eroded as a gulch in the high terrace in pre-Wisconsin time and its floor and walls included monzonite bedrock, Tertiary shales and sands, and pre-Wisconsin morainal material. During a pre-Wisconsin glacial advance, it was filled with outwash gravels that are largely composed of granitic boulders and pebbles, but included also considerable quartzite and gneiss.

This channel had sufficient gold concentrated at the base of the gravels to make it profitable to mine. The total area of the pit mined by Winkler is about 100,000 square yards, and between 400,000 and 500,000 cubic yards have been mined. This ground produced 2579.32 ounces of gold from 1913 to 1937. In the pit mined during 1938, at the north end of the area shown on Plate XI, this old channel had narrowed down and the gold content had diminished. About 12 feet of bouldery gravel, somewhat weathered, lay on partly decomposed granitic bedrock, and was overlain by 6 to 8 feet of arkosic sand. Water under a head of 125 feet was delivered to two giants, one of which was used for driving the gravel into the boxes, and the other to stack the tailings. Five lengths of 30-inch sluice boxes set on a grade of 9 inches to 12 feet were in use. The upper three boxes were provided with 2-inch by 6-inch plank riffles, and the lower two boxes with pole riffles. The results of the season's operations were unsatisfactory and preparations were made to mine in 1939 on the high bench west and some distance south of the 1938 pit.

As is shown on Plate XI, there is a large area of low-terrace and present stream gravels along lower Grouse Creek and between the old placer pits and Secesh River. This flat is known to be gold-bearing and is said to be from 30 to 40 feet deep to bedrock, but the upper 20 to 25 feet of the deposit consists of tailings from the Golden Rule placer workings, which would have to be handled if the ground were to be mined. The ground is physically suitable for dredging, but little is known of its average gold content.

Secesh Valley

The Secesh River Valley, from the mouth of Flat Creek to the lower end of the Meadows is a structural valley formed by block faulting, the main fault lying along the southwest side of the valley and the valley itself being faulted down in relation to the mountains to the southwest. There is also one known subsidiary fault in the valley itself, and there may be others that have not been recognized. The structural origin of the valley is plainly shown by the presence of southwest-dipping Tertiary beds exposed in the old Thorp placer pits, and Tertiary materials have been penetrated by drill holes at many places in the lower meadows. The geology of this valley is shown on Plate XII. Placer mining has been carried on in the past in several places in the Secesh Valley, and the meadows have been extensively prospected as possible dredging ground.

Thorp mine

The old Thorp placer pits lie on the southwest side of Secesh River and about half a mile south of the mouth of Grouse Creek. The property is said to have been first owned by Robert Royal, who mined it in a small way 50 years or more ago. Royal sold the ground to Toler and Thorp, who began mining on a larger scale about 1902 and continued for four or five years. During that period, Toler sold his interest to Thorp. The area exploited is a portion of a high terrace, the surface of which stands about 30 feet above the flat of Secesh River (Pl. XII). At some time in the distant past, the course of Secesh River lay to the southeast of the narrow canyon in which it is now confined near this ground, and, under the influence of an over load of gravel probably supplied by glaciers during a pre-Wisconsin Pleistocene ice advance, filled its valley with gravel to the level of the high terrace. The river then found a course to the northeast of the granitic hill that here restricts the valley, and during the time that elapsed before the Wisconsin ice advance had cut its canyon almost to its present depth, and removed much of the high-terrace gravel from its valley. These terrace gravels contain some placer gold and are to be correlated with the high-terrace gravels on lower Grouse Creek (Pl. XII).

Plate XII. Secesh Meadows, looking toward the canyon at their lower end.

The old pits on the Thorp property give striking proof of the structural origin of this part of the Secesh Valley. At the point where the road crosses the low ridge near the old Beaton cabins the hill to the northeast is composed of granitic rocks, but to the southwest the placer excavations show almost continuous exposures of Tertiary sand, shale, and lignite that dip to the southwest, there to terminate abruptly against granitic rocks along a fault. The section shows a stratigraphic thickness of at least 650 feet of Tertiary beds; therefore, the displacement along the fault must be at least that much and may be much more (Pl.III). In the old placer pits, the bedrock everywhere consists of Tertiary sedimentary rocks, and exposures in the old workings of the Golden Rule mine on Grouse Creek, together with many bore holes in the Secesh Meadows, indicate that Tertiary beds are widely present in this basin beneath the younger gravels.

The old workings were mined by hydraulic methods, water under head being brought from a small tributary creek that joins Secesh River from the southwest. From 5 to 12 feet of terrace gravels were sluiced and the cuts were extended several feet into the fairly soft Tertiary bedrock. The gravels contained a few boulders as large as 2 feet in diameter, many of head size, and abundant cobbles from 4 to 8 inches in diameter. About 80 per cent of the gravels consist of granitic rocks; the remainder consist mainly of quartzite, although some cobbles of vein quartz, pegmatite, and gneiss are present. Many of the granitic boulders and cobbles are hard and firm, particularly the finer grained varieties, but many are decomposed and many that went through the sluice boxes intact are now crumbling on the dump. In places, particularly in the southernmost pit, the surface of the gravels was covered by as much as 4 feet of impure peat overlying well rounded gravel. Practically all the gravel was put through the sluice boxes, and very few boulders were stacked in the cut. This high-terrace gravel is believed to be outwash deposited during an ice advance that intervened between the Wisconsin glaciation.

It is reported that, although a large amount of ground was mined on this property and considerable gold was recovered, nevertheless, the operation was not profitable and no mining has been done since about 1905. No information was available as to the average gold content of these gravels.

Gayhart Burns mine

A group of claims on the east side of Secesh Valley, opposite the Fernan ranch, was staked in the early days of the camp by Gayhart Burns and was mined by him until 1905. This property is known locally as the Gayhart mine. Both to the north and to the south of the ground mined, the deeply weathered granitic bedrock crops out as cliffs and residual boulders, but at the mine there is a deposit of unconsolidated material in which all of the boulders on the surface were composed of quartzite. Excellent exposures on the property disclose the fact that this deposit is an ancient moraine, probably of early Pleistocene age. This moraine originally consisted mainly of coarse, angular to partly rounded, granitic boulders, many of which were from 3 to 8 feet in diameter, with small amounts of gravel of other materials in the interstices between the boulders (Pl. V). It seems probable that this morainal material was distributed along the valley slopes of Secesh River as far south as this point, but has since been eroded away at most places. In the long period that has elapsed since this moraine was deposited, it has become deeply weathered; the granitic boulders have lost their coherence, and, although still recognizable as boulders in fresh exposures, they are actually only aggregates of arkosic sand. The quartzite boulders by contrast are still hard and firm. In addition to the deep decomposition of the moraine, surface creep and erosion have destroyed the morainal character of the surface features, and the presence of a body of morainal material can be recognized only by the quartzite boulders that are scattered over its surface. Apparently, the area in this vicinity in which the old morainal material is present is small.

Exposures in the old placer pits show that a surface layer of gravelly material from 6 to more than 10 feet thick was mined, the floor of the cut in places consists of old moraine, and elsewhere of deeply disintegrated granitic bedrock. Recent gullies in the old moraine show the granitic boulders to be thoroughly disintegrated to depths of 25 feet below the floor of the pit, and, therefore, 30 to 35 feet deep below the original surface before mining. No information is available as to the gold content of this ground per cubic yard, or of the total amount recovered, but it was enough to encourage Burns to continue mining for 35 years, to the time of his death.

Secesh Meadows

The name of Secesh Meadows, as used here, denotes both the flat valley floor of Secesh River that extends from a point half a mile below Long Gulch to the mouth of the canyon above the mouth of Flat Creek and the considerable area in the lower valley of Grouse Creek. In this stretch, a distance of about 5 miles, the valley floor is occupied by meadow lands and low terraces that stand from 4 to 8 feet above the stream at its summer stage. The meadows are divided into two parts by a constriction in the valley at the old Thorp placer mine. The upper portion forms a triangular area that (Pl. XIII) is about 6,500 feet long, parallel with Secesh River, and extends northward 4,000 feet into the lower valley of Grouse Creek. This ground is said to be owned in part by Fred and Alfred Clark of New Meadows, and in part by L. E. Winkler of Council, Idaho. Its total area is somewhat less than a square mile. Below the Thorp mine, the lower meadow is about 3 miles long and has an average width of 2,000 feet. It is for the most part in grass and free of timber, and is physically well suited to dredging. Four claims in this meadow are owned by R. A. Crocker of Spokane, Washington. Both the upper and lower meadows have been extensively prospected by pits and by drilling, but the results of much of that work are not available. R. A. Crocker generously furnished a blue print of the ground, and many profiles across the valley that are given herewith (Fig. 3). The information he supplied was obtained from various sources and has not been verified by the writer. Others who have done some prospecting of this ground report somewhat lower gold content than that given in Figure 3. Moreover, as the bedrock in many holes consisted either of soft Tertiary beds in place, or of thoroughly decomposed granitic material, it became a question of the driller's judgment as to when the alluvial material had been passed through and undisturbed material in place had been reached. The profiles in Figure 3 reflect that judgment, and might be modified considerably in the light of actual mining. Apparently, the ground is not of high-grade, but it contains enough gold to merit consideration, and with the proper equipment and prudent management might be exploited at a profit. In the lower meadows, the depth to bedrock is as much as 25 feet and in the upper meadows as much as 30 or 35 feet of bedrock is in part decomposed granite, and in part rather soft Tertiary clays, sands, and lignite, into any of which a dredge could dig deeply enough to affect a thorough cleaning of bedrock.

Summary of placer possibilities

The placer diggings of the Secesh Basin have never been exceptionally rich, and, although considerable ground has been mined at a number of places, only those enterprises have been successful that have had skillful and thrifty management. It is fair to assume also that those areas that have been mined were selected because they offered the most favorable ratio between the cost of mining and the value of the contained gold at the time that mining was undertaken. However, since the early days of the camp, conditions have changed greatly in a number of respects. The accessibility of this basin has been greatly improved by the opening of roads so that freight costs are only a fraction of what they were 40 years ago; mechanical appliances for handling placer ground have been developed or perfected, including the introduction of the dragline excavator and the bulldozer; the efficiency of the gold dredge has been increased, and the price of gold has been raised from $20.67 to $35.00 an ounce. Of particular importance, in this area of low stream discharge in the late summer and fall, is the development of the dragline scraper and movable washing plant in which the water is recirculated by pumps so that only a small supply is necessary. All of these changes have resulted in lower costs of moving and sluicing placer ground, and have made possible the profitable exploitation of ground that could not be mined at a profit under the conditions that prevailed a generation ago.

The Secesh Basin still contains large bodies of gold-bearing placer ground, some of which will certainly be mined in the future and others of which deserve thorough prospecting to determine whether or not they are minable under present conditions. The high terraces along the east side of Lake Creek, from the upper forks down to and somewhat below Three Mile Creek, have been mined in places in the past and locally may contain enough gold to be profitably worked at present. There is a large body of this material that is physically well adapted to mining by hydraulic methods or by dragline. Above this terrace on the valley slope, and extending to a height of 300 feet or more above Lake Creek level, there are extensive deposits of old morainal material, the presence of which is indicated by numerous quartzite boulders and cobbles on its surface. In this material, most of the granitic boulders are so decomposed that they disintegrate upon mining, and the quartzite boulders are neither too large nor too numerous to be a serious deterrent to mining. This old moraine is gold-bearing, has been mined locally, and in places may be rich enough to be profitable. It should be more thoroughly prospected. Lake Creek, from its forks to its mouth, flows through a series of flat meadows that are known to contain placer gold. Such meager prospecting as has been done indicates that these meadows are of low-grade, but locally may contain minable ground. They have never been adequately prospected with possibilities as dredging ground in view.

The valley of Secesh River, from the mouth of Ruby Creek to a point about 2 miles upstream, is floored with gravels that represent outwash from the last glaciers. This flat contains abundant large granitic boulders, and is said to contain only a small amount of gold. It would be difficult to mine and is not believed to be potential placer ground.

The low ridge that lies between lower Ruby Creek and Secesh River contains extensive deposits of morainal material of two ages as well as remnants of old stream gravels that locally underlie the morainal deposits. Both moraines and the underlying gravels contain placer gold, and have been extensively mined in past years. A rough estimate indicates that there is a deposit containing 5,000,000 to 10,000,000 cubic yards of this material, and mining has shown that at least part of it contains from 15 cents to 20 cents a cubic yard in gold. Some of the old gravels that lie beneath the youngest moraine are very rich. There is a possibility that with an adequate and continuous water supply much of this deposit could be profitably mined.

Extensive meadows in the upper basin of Ruby Creek contain some placer gold and are physically adapted to dredging. Such meager information as is available concerning their gold content indicates that they are of too low grade to be attractive at present.

In the lower valley of Grouse Creek, and the adjacent portion of the Secesh Valley, there is an area of more than a square mile of terrace and stream-flat gravels that are gold-bearing. The terrace gravels have been mined at several places. Part of this ground is physically well adapted to hydraulic mining, and the flats are suitable for dredging. This area contains many millions of yards of gravel, some of which certainly will be mined in the future. Little information is available as to its average gold content.

The lower Secesh Meadows have an area of about 3,500,000 square yards. Assuming an average depth of 15 feet, there is apparently about 17,500,000 cubic yards of material, the average gold content of which is said to be between 10 and 15 cents a cubic yard. This ground will doubtless be dredged some day, although, so far as is known to the writer, there are no immediate plans under way to develop it.

To summarize, there are large volumes of gold-bearing gravel in the district, but most of them are of such grade that efficient methods and prudent management will be required if they are to be profitably exploited. No extensive mining enterprise should be begun without thorough prospecting of the ground.

www.ingramcontent.com/pod-product-compliance
Lightning Source LLC
Chambersburg PA
CBHW070811210326
41520CB00011B/1904